VIEW FROM THE COCKPIT: LOOKING UP

Archie B. Lawson

Archie B. Lawson
Riverton, Utah

Veterans Breakfast
High School
Sneedville, TN.
May 31, 2008

Unless otherwise indicated all Scripture quotations are taken from the New King James Version®. Copyright © 1982 by Thomas Nelson, Inc. Used by permission. All rights reserved.

All the stories in this book are true. They are my experiences covering most of my life.

Cover designed by Archie B. Lawson. The sunset background was taken on an arrival into Salt Lake City International Airport. The circle cockpit was taken of a Boeing 757 parked at the gate before departure, and the photo of the airplanes seen out the front window was taken on the taxiway at Seattle International Airport.

Black and white photos of all the different types of airplanes I flew professionally are in chronological order and appear with a brief description.

View From the Cockpit: Looking Up

Published by Archie B. Lawson

Library of Congress Cataloging-in-Publication Data

Lawson, Archie B.
View from the cockpit: looking up/Archie B. Lawson

ISBN 0-9762660-0-8

Printed in the United States of America

Acknowledgements

I want to thank

- My wife Jody for her many years of support and encouragement. Without her loving nudging, this book would not have been written. She often reminded me of the importance of these stories.

- Leigh Prim of Southeast Baptist Church, Salt Lake City, Utah, for countless hours of competent assistance in the important task of editing. She has truly been a blessing to me through the service she so gracefully rendered.

- Robert R. Zeimer Jr. for his dedicated assistance in proofreading this manuscript.

- All the people who have been in my Sunday school classes over the years who have told me how much these stories have meant to them.

- God for giving me my wife Jody, friends like Leigh Prim, my brother-in-law Robert R. Zeimer Jr., and all those who have encouraged me throughout my life. I also thank God for giving me a job that was a joy throughout my entire career.

This book is dedicated to my wife, Jody, for
praying for me as I traveled over ten million miles.
Her prayers paved the way for me to share the
gospel with many of my crewmembers.

Piper J3 Cub

When I was six years old I went for my first flight in a Piper Cub such as the one in this photo. The primitive cockpit is pictured below.

Contents

Beechcraft T-34B Mentor

I began my aviation career in this type of airplane in Pensacola, Florida. My first solo flight was in a T-34, and I performed numerous aerobatic maneuvers as part of my training. The Navy bought 423 of these aircraft from Beech. Most of these airplanes were flying out of a satellite field of NAS (Naval Air Station) Pensacola and from NAS Whiting Field, Florida. The Navy used the T-34B for over twenty years accumulating around 100,000 flight hours per year. One aircraft logged over 5,000 airframe hours which included over 16,000 landings, almost 18,000 stalls, and over 4,500 loops.

Preface

> I press toward the mark for the prize of the high calling of God in Christ Jesus.
>
> Philippians 3:14 (KJV)

Why has man dreamed of flying through the centuries? What stimulated him to dream the magnificent feat of soaring through the heavens? Was he seeking to physically elevate himself or did he have sacred reasons? Perhaps he wanted to get closer to the gods, or the God, that he worshipped. I have read that man's desire to emulate birds may have been the result of envy, the oldest deadly sin. Whatever the motive, this fascination exists for many today. A number of people have told me over the years that they wished they had learned to fly. Many of these people regretted not pursuing a career in flying while others only wished they had learned to fly smaller airplanes. I will relate some of the memorable stories associated with my career in aviation, as it was a delightful ride and I want to share some of it with you.

One of the best known legends about flying is that of Daedalus, the engineer who built the labyrinth on the island of Crete in which the Minotaur lived. With his son Icarus, he was imprisoned by King Minos, but the captives escaped by making themselves wings of wax and feathers. Daedalus flew successfully all the way to Naples. However, Icarus, excited about the thrill of the new experience of flying, let his youthful exuberance deafen him to his father's warnings, and he flew too near the sun. This caused the wax to melt and sent the boy crashing to his death in the sea below. As you proceed through this book, you will see that it is extremely important to heed the counsel of our heavenly Father when it comes to flying and living life from day-to-day. God uses many avenues to issue these warnings. If you heed His warnings, your flight through life will be filled with greater joy and contentment.

Fasten your seatbelts and come with me for a look back at some of the experiences that made my career so enjoyable. As I tried my very best on all of my flights, I will try to give you a smooth taxi, takeoff, cruise, and landing. I trust that some of these stories will inspire you in your place of work or wherever God has placed you. My goal for you is that you take the experiences from your vocation, along with other areas of your life, and apply them in such a way as to make the Kingdom of God come alive as you share the Gospel with the people you see on a daily basis. All you have to do is join Him in His work wherever He places you. It is my prayer that God will greatly bless you as we take this flight together.

I pray that you will do more than just read and enjoy some of my stories. It is my desire for you to be able to use your own examples for the glory of God. This book is the result of many years of urging and encouragement from my lovely wife, Jody. Many of the stories are from notes she took when I was teaching Sunday school or teaching in connection with our volunteer financial counseling. I have made it a practice to use examples of my flying experiences in my teaching. She would also record incidents I mentioned during the discussions we had upon my return from my many trips. Jody sat and logged notes of my experiences with the hope of someday getting me to put them down on paper. When she presented me with a copy of these notes many months ago, I was both amazed and shocked at the same time because I had no idea she was recording all these stories. If you like these accounts, you have Jody to thank for never giving up on me to get this job done.

One of the things you will see as you read this book is my love of flying. The main reason I see for keeping this attitude throughout my career is that I kept my eye on the prize. My favorite parts of the physical act of flying were the takeoffs and landings. Out of these two, I especially looked forward to every landing. I viewed the runway at the airport as the prize, and kept my focus and pressed toward the mark for that prize. There is another prize on which I am focused, and it is far superior to the one mentioned above. One of these days I will make my most spectacular landing where I will claim the ultimate prize.

North American T-28B Trojan

This was my second type of training airplane. In addition to learning how to fly formation, I performed countless aerobatic maneuvers. The Navy flew the version designated T-28B. The "B" came equipped with the powerful Wright Cyclone R-1820, 1425 hp engine, Hamilton Standard 3-blade propeller, belly mounted speed brake, and many other changes and improvements. 489 "B" models were built and used from the mid-1950s to the mid-1980s. In addition to the "B" model, the T-28C was built for the Navy starting in 1955. The T-28C is equipped with a tail hook, a smaller diameter propeller, and other small changes to allow aircraft carrier landings. 299 "C" models were manufactured with production ending in 1957. I made my first carrier landings in the T-28C.

My View From the Cockpit

> And God called the dry land Earth, and the gathering together of the waters He called Seas. And God saw that it was good.
>
> Genesis 1:10

My view from the cockpit was much more than a view of the physical beauty I saw through my front window. That, in itself, was often breathtaking. One day I would fly over the Grand Canyon and enjoy the great expanse of lavish scenery for many moments. The next day I would gaze at the Statue of Liberty or the Empire State Building on my approach to an airport in the New York City area. I have been blessed to see the Northern Lights, the glaciers of Alaska, the mighty Mississippi River, the Golden Gate Bridge, and many significant mountain ranges in the world. In my travels I have seen almost every conceivable cloud formation, and I have witnessed the beauty and devastation nature can unleash. Having Jesus Christ in my life has allowed me to see much of the world's physical beauty from an enhanced perspective. I want to focus on the beauty of knowing Jesus Christ as my Lord and Savior.

One of my primary goals in flying was to be open to the opportunities God gave me to share with others about the most important subject in the world, Jesus Christ. I would pray for unique opportunities to share the Gospel. I was often astounded with some of the examples God gave me to use. Here is just a small sampling of a common occurrence where God allowed me to use a spiritual principle.

A flight attendant brought some honey roasted peanuts to the cockpit. The copilot quickly ate his, and I opened mine and began to enjoy them. I offered to share some of my peanuts with him, but he declined because he hadn't shared his with me. I politely reminded him that the love that Jesus had for me was not dependent on whether I shared my honey roasted peanuts with Him. His love for me carried no conditions. With these statements, he accepted some of my snack. I loved those wonderful tasting nuts, but I was mindful of the fact that standing up for Christ often requires a price. The price in this example was that I would have to eat a smaller amount, but I assure you that I never had any regrets in paying this price. If we choose to hold on to things too tightly, we might miss an opportunity to share something good with someone else. I know this is stretching the point somewhat, but the principle exists.

My desire is to attach spiritual significance to the experiences I relate to you. I use examples of everyday flying situations from my life and apply spiritual principles to them. I used this method when sharing the Gospel with my coworkers. I pray that you will be encouraged to share the Word in a practical way with whom you come into contact.

My dream began when I was six years old. My dad and I were on our way to the hospital in Greenville, Tennessee, where I was scheduled to have my tonsils removed. On the way to the hospital, we passed a highly groomed farmer's field. Next to the barn, where one might expect to see a tractor, there was a J3 Piper Cub airplane. Dad stopped the car so we could take a closer look. During our brief examination of this strange craft, the farmer appeared. We exchanged greetings, and he asked if we would be interested in going for a ride. I believe dad gave him a couple of dollars, and we were treated to a ride that would have a life changing impact on my future. I knew that I wanted to learn to fly from that moment, but never considered making it my vocation. All I really wanted to do was find another J3 Piper and go for another spin, again and again.

My career began in July of 1962. Upon graduation from Carson-Newman College in Jefferson City, Tennessee, I set out for Pensacola, Florida, to embark on a career as a Naval Aviator. My intention was to fly for the Navy for about 20 years and then become a public school teacher. After five years in the Navy, I decided to leave the military and pursue a career in commercial aviation because I could continue to fly beyond the twenty year mark. In commercial aviation, a person can work for a major air carrier until the age of sixty, the mandatory retirement age by the Federal Aviation Agency (FAA). There were numerous factors that influenced my decision, the main one being that I simply loved manipulating the controls of the airplane. There was no questioning the fact that I cherished the ***view from the cockpit***.

My thinking changed from wanting to be a part-time pilot to having a chance to fly for a profession during my sophomore year at Carson-Newman. I went to the library to do some research for a book report and saw a Navy recruiter sitting in the middle of the library in a full-dress officer's uniform, complete with the shiny golden wings of a Naval Aviator. Next to him was a sign that asked the question, "Have you ever considered a career as a pilot for the Navy?" The answer to that question for me was a big fat "NO." I only went back to speak with him because I knew he was familiar with airplanes. I thought it would be an interesting conversation because of my interest in flying. During our little chat, he asked me if I would be interested in taking a qualification test. If I agreed to take the test and passed, I would be flown to Atlanta, Georgia, to take the full test. I only decided to take the exam because there was a free plane ride for passing.

Well, I took the test and passed. A short time later I found myself in Georgia taking the full-blown test to have a chance to become a Naval Aviator. To my surprise and astonishment, I passed the written and physical exams but there were still a couple of problems. I was only 18 years old, and one must be nineteen and one-half years old before entering the Navy flight program. The other problem was that the sophomore year of college must be completed and I was still in my sophomore year. I was told to let them know when either of these things occurred. Since I was so close to finishing college, I made the decision to get my degree before going into the Navy, and completed my college studies in three years with a Bachelor of Science in Mathematics. This is a decision that I have never regretted because it gave me the opportunity to go through the flight training phase as an officer. Without the degree I would not have been commissioned until completion of training. As we progress, I will relay some of my experiences during this fascinating time of my life.

My mother was not pleased when she heard that her son was going to take an airplane ride to Atlanta for a chance to fly as a career. She had never been on an airplane because she considered it to be too dangerous. I never wanted to go against my mother's wishes, so I had a dilemma. This was solved by my roommate in college at the time. He was a four year Navy veteran, and he spoke with my mother and convinced her that there was no way I could ever pass the physical exam that the Navy required. With this assurance, my mother reluctantly agreed to let me go. After my trip to Atlanta, she knew that I was planning to do something that would cause her much worry and concern in the years to come. Many years later, my mother had an opportunity to fly on one of my flights from Atlanta to Salt Lake City, Utah. We encountered some very rough thunderstorms, and the flight was rather bumpy. I was concerned that she was not enjoying herself because of all the turbulence. After we arrived in Salt Lake City, I asked how she enjoyed the flight. She told me that there was never a concern because I was the captain of the flight. My dad never had the opportunity to fly with me on one of my commercial flights.

In my discussions with people over the years, I have found that the same questions are asked. Have you ever had any close calls? Have you ever had any famous people on your airplane? What is it like to be responsible for all the lives in your hand? Do you ever get scared? Do you ever get tired of all that traveling? The answers are yes, yes, God is in control, no, no. You will read about some of these things as we go along. This will include my worst close call, the title of which is, "My Fourteen Year Close Call."

On August 21, 1967, one month after being discharged from the Navy, I began my civilian aviation career with Western Airlines. I was based in San Francisco for all of my time with them except for my initial training in Los Angeles, which was the home base for Western. In 1987, Delta Air Lines merged with Western. My first choice for employment was Delta because they were based in Atlanta, and I grew up in the South. This would have given me the opportunity to live closer to home. I was not hired by them because they were one of six airlines that had a minimum height requirement of five foot eight inches, at that time, on their application. The height limitations were removed in the early 1970s when commercial airlines started to accept women pilots. This did not turn out to be a negative thing because I never would have met my wife, Jody, if I had started my career with Delta in the South. I had the opportunity to work for these two great airlines. From the time of the merger in 1987 until my retirement in 2001, I was blessed to be associated with the best airline in the world.

Grumman S2F Tracker

This is the type of aircraft I flew in advanced training at NAS Corpus Christi, Texas. The Navy relied heavily on the Tracker for fleet defense as an antisubmarine aircraft. 200 S2F-1s were converted into "S2F-1T" trainers for multiengine flight training. In addition to my regular training, I also had eight carrier landings aboard the U.S.S. Lexington in the S2F. The nickname for this plane was "Stoof." This was a play on words for S2F.

Growing Up on the Farm

My birth certificate says that I was born in Sneedville, Tennessee. At the time, the population was only a few hundred people. My parents lived in a small wood frame home on the farm belonging to my paternal grandparents, Charles and Bertha Lawson. Sometime after my first year, we moved to the farm where my maternal grandmother lived. Our home was a small five room house with two bedrooms and the dining room was later turned into a bedroom. The house was heated by a coal burning stove in the corner of the living room. The toilet facilities were not inside the building. In the early 1950s, my dad had a well dug, and, for the first time, we had running water inside the house. We still did not have an indoor bathroom. My first full-time experience with an indoor toilet was when I went to college at the age of seventeen. This was a rather pleasant change.

The house we lived in was built primarily by my uncle Darryl. He passed away when I was too young to remember him. Mom also had a sister, Geneva, who died before reaching the age of twenty. My grandfather, a World War I veteran, died when mother was very young. My grandmother married George Rhea several years later, and she lived until 1949. After her death, our family inherited a small fifty acre farm. The primary money crop was tobacco.

Beginning in the mid-1940s, dad started working at Harrison's Hardware. He ran that store until he began having medical problems in his mid-60s. This is when he suffered his first of several heart attacks. He passed away in 1993 from prostate cancer. Mother had the primary responsibility of raising me and my two brothers because dad would be at the store from very early till after dark. This was his ritual six days each week all year long. This little store carried a broad range of merchandise such as appliances, paint, televisions, radios, nails, horse collars, floor coverings, bedding, and just about anything needed around the house. If it was something people really needed, it could be found there. One would have to travel to Morristown, thirty five miles away, to satisfy some wants. My dad sold the first television in Hancock County. Our family purchased a set a few months before you could watch any programs, because there were no stations in the area. The only thing we could see during those months was a snowy test pattern from somewhere in West Virginia, because it was not until 1954 when the first station came to Knoxville.

On Saturdays, I often had the chance to go to town and see the "big city," the largest town in the county, with a population of seven hundred and fifty people. In addition to the visits to dad's store, I always looked forward to going to Turner's Drug Store. They made the very best lemonade that I have ever tasted, and I never learned their secret. If I wasn't at Turner's having lemonade, I was down at the concrete block wall around the courthouse where people would come to sit and talk. The people there could be counted upon to have a pocket knife, and almost everyone would be whittling on a small piece of cedar. They weren't carving anything in particular, but they were just accumulating a pile of wood shavings as they talked. The discussions ranged from their latest crops, their families, animals, what they liked, or didn't like, and about the church service the previous Sunday. If one liked the sport of gossip, this was the place to be. This was before I knew what gossip was. You could hear the latest news about who was in jail, and you knew all the rumors about those still making moonshine.

For the first five years of my schooling, I attended a one room school called Duck Creek Elementary. The teacher had the responsibility for grades one through eight. After my first five years, the county closed this school. We had an outdoor basketball court that was a flat dirt area. There was also an outdoor toilet and a wonderful place to play marbles. The school was in a wooded area, and there was no shortage of trees to play in or on. It was in those trees where I did my best to imitate Tarzan. On one of my jumps in the early morning before school began, the bark gave away, and I fell. This fall left me with a broken arm, and I remained at school the entire day. When I arrived home that afternoon and my mom saw me holding my arm, she knew right away that it was broken. We did not have a phone to reach dad at work, so mother had to summon a neighbor about a quarter of a mile away to take me to Sneedville for treatment. The doctor there said the break was too difficult for him to treat, and told my father to take me to Knoxville which was seventy five miles away. We did not return home until well past midnight. My days of wanting to be like Tarzan came to a halt. I was nine years old when this accident happened.

The cost of living was very low where we lived. Hancock County was one of the five poorest counties in the United States. We may have been poor, but we always ate well. Mom raised a big garden, which would sustain us for the entire year. During the non-growing season, we enjoyed wonderful canned and frozen food from her garden. We raised our own hogs and chickens. I don't ever remember when we didn't have plenty to eat. If we had anything that required refrigeration, we had to place it in the water in a small creek house. This was a small structure over a cool running stream. Containers of milk would be stored there until we were ready to use it. Anyone who had a refrigerator was considered to be well off. This little stream was located a good distance from our home, and we got our exercise going for a glass of milk.

Our farm had too many hills for a tractor to be used. We didn't have the resources to purchase one even if this wasn't the situation. Heavy farm chores were primarily done with mules used for cutting and hauling hay as well as plowing all of the fields. You may have seen some of these mules in your history books, or a documentary about early life in America. Living in that part of Tennessee was comparable to life many years earlier. One had to be in good physical condition to run a farm. This is one of the main reasons that I decided at a young age, about eight or nine, that I would try to do anything within reason to get away from the farm. I never viewed this as a lower class of life. I just recognized that I could not meet the physical requirements. During the more difficult time of my schooling, I would often revisit the thoughts I had at that young age. This gave me that extra measure of determination to finish the job. I don't know if I could have accomplished the task before me without those memories. Today, I see that time of my life where God was preparing me for His use at a later time. I could never have done it without Him.

After Duck Creek Elementary closed its doors for the final time, I went to a four room school at Greene-Lawson Elementary. For the last three years of my pre-high school years, I was in a classroom with only three grades. This is where I completed the sixth, seventh, and eighth grades. Upon finishing the eighth grade, I was then off to the big time: I was on my way to high school. Hancock High was the only high school in the county. As I got older, there is no doubt that God was orchestrating my life. Early in my time there, I made friends with four other guys. Their names are Wendell Trent, Stanley Trent, George Davis, and Douglas Seal. Douglas passed away in 1989 at an early age from a stroke, and Stanley died early in 2004 from cancer. When I return to Tennessee, I still get together with the other two.

These friends with whom I associated were serious about their education, and they had somewhat of a vision for their future. My vision was limited compared to theirs, but some of their dreams rubbed off on me. I never cease to thank God for giving me their acquaintance because of the influence they had on my life. Throughout our four years of high school, we were always competing to see who could get the best grades. Douglas was the valedictorian, and George was salutatorian of our graduating class. I always brought up the rear. They always excelled in our quest for the best grades, but I never gave up trying. Of the people in our class, we were the only men who went on to graduate from college. Since the foundation was laid for our education before college, we had something on which to build. I don't think the others needed me, but I sure needed them, and I don't believe I could have done it without their influence on me. We all left high school and pursued our continuing education at different schools.

My father did not work on Sundays, and that was always reserved for attending church, primarily at Duck Creek. After church, we would take an occasional trip to Kingsport to see relatives. Four of my aunts lived there with their husbands and children. This was always a trip to which I looked forward. It was a welcomed break from the farm life, although I never remember this life being terrible. All I remember about the negatives was how difficult the work was on the farm. There are many pleasant memories of that simple life. When I have been faced with great amounts of stress, I confess that there were times when I longed to return to the simplicity I experienced at that time.

Until my grandfather on my father's side passed away in 1956, we often spent many Sundays at their farm as well. My grandmother lived until 1982 when she died at the age of eighty-nine. It was a common thing for much of the family to gather there where we could count on a wonderful feast prepared by my grandma. She is the only one that I ever knew who made gooseberry pie, and their farm was the only place where I saw gooseberries growing. Many years later, I found some gooseberry jam at a specialty story in Seattle. My uncle Grover lived on the farm where my father grew up until mid-2004. It is great to reflect on these precious moments.

In the early 1980s, I was in the town of Morristown, Tennessee, visiting with my dad's good friend, Elmer Greene. He built the home my mom lives in today. While I was there he told me about an electrical job he did for my grandparents in the early 1940s. He had wired their home for electricity, charging five dollars for the work. My grandfather gave him a mantel clock to satisfy the bill. I asked him if he knew where that clock was, and he said that he had given it to his brother in Knoxville. My curiosity was becoming a little active by then, and I wanted to know if there was any way I could obtain the clock. About two weeks after I returned to my home in California, I received a package in the mail. The package contained this clock. I still have it above the fireplace in our home. Every time I walk by that clock, I am reminded of the pleasant times I had with those grandparents. The farm was sold in 2004.

The house I grew up in was torn down over thirty years ago. A few years ago, I did a simple pencil drawing of that farm house where I lived until going away to college in 1959. This drawing was done as a gift for my mother.

My Life On the Farm

Living on a farm in the deep hills of Tennessee,
Gave me that wonderful sense of being free.
I could roam the fields of clover and play in the hay;
There was never a shortage of things to fill the day.

I could take a leisurely walk through the woods,
Or I could go out and harvest nature's goods.
The cows needed to be milked and the animals fed,
By the end of the day I was ready for the bed.

At night as I said my prayers and went to sleep,
I always slept soundly without any peep.
There were no cares or concerns to keep me awake,
I didn't have to worry about having all I could take.

There was no worry about the bills I couldn't pay,
I knew I always had a great place to stay.
My mom and dad provided the best they could,
I never gave them all the credit that I should.

My dad worked hard, and my mom gave her all,
They didn't complain, they were always on call.
I thank God for giving me parents who cared so much,
My mom was there when I needed that special touch.

If I had a chance to again walk right through that door,
Would I like to go back and do it all over once more?
Oh, I don't dwell very long on what has long since past,
But I try to focus on what I can do that will always last.

A.B. Lawson, May 10, 2003

Lockheed C-130E

Upon completion of flight training in Corpus Christi, Texas, I flew the Lockheed C-130E aircraft for three years and accumulated over two thousand flight hours. This included thirty missions to Vietnam. Widely know for its versatility, the C-130 has been one of the most successful aircraft designs in history. The first model was built in the 1950s and is still being manufactured today.

God is in Control

> The steps of a good man are ordered by the LORD, and He delights in his way. Though he fall, he shall not be utterly cast down; For the LORD upholds him with His hand.
>
> Psalm 37:23-24

I want to make it perfectly clear that I am not using the above passage to insinuate that I am, or was, a "good man." My purpose is to show how God had His hand on my life when I was on the outside looking in. There is no question that He was in full control, but that did not stop me from trying to run things on my own. I give praise to the One who thwarted many of my puny efforts to make things work "my way." Some of the things I am sharing with you have never been shared with anyone, and the only reason for doing so now is to encourage others to take a look at themselves to see how God "orders our steps." *A man's heart plans his way, but the LORD directs his steps.* Proverbs 16:9 If we can fully grasp God's work in our lives, we can be much more effective in our witness of Him to those we encounter on a daily basis. I listened to two different voices while I considered the stories in this chapter. One voice told me to do what I could to make myself look as good as I could, and another to reveal the fact that I am less than perfect.

I observed a dramatic experience from the airport in Knoxville that could have ended my quest to become a pilot. I was on my way to Atlanta to take the physical and written tests for the Naval Aviator Program. As I waited for my flight, I had the opportunity to see many different types of airplanes. I had previously only seen these planes in pictures, and I felt like a kid in a candy store. Since I was going to be flying to Georgia on a military DC-3, I was on the opposite side of the airport from the civilian commercial aviation terminal. One of the types of airplanes stationed there was the F-104. At the time, this was the fastest fighter in the arsenal of the U.S. Military. As I stood on the ramp gazing intently at all the aircraft, I had the pleasure of watching an F-104 taxi out for takeoff. The sleek and trim machine continued onto the runway, the sound deafening as the pilot added power to leap into the sky. I stood in awe of what I observed.

Little did I know that this would be short lived! When the F-104 lifted off the runway, it disappeared from view over a hill at the perimeter of the airport boundary. There were several people with me intently waiting for the shiny airframe to come back into view, but this was not going to happen. We did see the top of a parachute just above the ridge line, but no plane. Shortly after seeing the parachute, we saw a black cloud of smoke coming from the same general area. It did not take us long to learn that the airplane had crashed on takeoff, and never reached an altitude of more than one hundred feet. The pilot activated the ejection seat which sent him upward about three hundred feet. The reason for this is that you can eject from a crippled plane near the ground, and there will be sufficient time for the chute to open. One of the people standing beside me on the ramp was the crew chief of the downed craft, and he almost went into a state of hysteria because he was the one responsible for the maintenance on that particular machine.

How did this affect my thinking about wanting to fly? I did not contemplate abandoning my quest to reach for the sky. As I have relayed this story over the years, many have told me that this would have ended their dream. I don't really have an explanation for why I was not overly concerned about my future safety, but I believe it was part of God already knowing all the aspects of what was to come in my life while placing His arm of protection around me and my emotions.

After taking all the tests for the Navy, and passing them, I didn't stop there. Over the next few months, I would have many enticements thrown before me, and I must say that I gave many of them a closer look. There was a chance to go into the Marine flight program. I pursued this because of a program where I could receive my commission as an officer upon graduation from college if I signed up for summer active duty assignments. I wouldn't have to wait four months as in the Navy arrangement. Since I was into the "living for the moment" way of life, I applied to take advantage of this offer. This meant that I would get another free plane ride to Atlanta to take the tests again. The tests were given at the same place as the previous ones because the Navy and Marines go through much of their initial flight training together. I passed all of the written tests, but I failed the physical because my weight was below their standard of one hundred thirty-five pounds and I weighed one hundred twenty pounds. They said they wanted men, not boys. In retrospect, I am very thankful that I didn't qualify for the Marines. The type of active duty would have been drastically different, and would have exposed me to much greater danger. When we take things into our own hands, we often expose ourselves to circumstances and situations that could have detrimental results. God was again in full control of my life.

After I was turned down for the Marines, someone told me about all the attractive features of the U.S. Air Force flight training. Once again, I tried to take things in my own hands, and I applied to enter their program. This little detour lasted only a short time. Just before I was to go to take the tests, I became ill. When I took the tests, I could hardly concentrate on the questions before me. I did not pass their written tests, and my attempt to do things on my own failed again. Failing the tests was a blow to my ego at that time, but I have given thanks to God many times for letting me fall short of meeting the minimum requirements. In the Navy, I was able to do things that I would never have had the opportunity to do in the U.S. Air Force such as landing an airplane on an aircraft carrier. I see this episode in the light of God placing me where He wanted me and receiving many benefits over those offered in the U.S. Air Force.

After completing the initial part of my flight training, I had the opportunity to choose the type of flying I wanted. The very best assignments were allocated according to the grades received during training. We were graded on flying ability, physical accomplishments, and military savvy. One of my friends decided he wanted to fly helicopters, and he was accepted into that program. This man had a way with words, and he presented me with his explanation of why he chose this route. After listening to his presentation, I thought I would enjoy some of the things of which he spoke. About a week later I decided to apply to fly those funny things that had the propeller coming out of the top. The day before I made my request, all the slots had been filled, and I was told it was too late. The more I pondered this, the more gratified I was for the way things turned out. That type of flying would not have given me the enjoyment I received flying fixed wing aircraft. This is not to say that the other is worse, but it was not the best for me. God allowed me to be in the right place for my demeanor and temperament. As I look back on some of these events, I believe they were all part of my elusive search to find something that was just a little bit better until I found the One thing that would fill the emptiness in my life. God was there to pick me up each time I fell.

Still, there were other times when I tried to make things just a little better. When I contemplated leaving the military for a job in commercial aviation, I applied to a number of airlines. My first choice was Delta Air Lines, but I didn't get past the application because they had a height requirement. After looking at a few other prospects, I applied to Western Air Lines in Los Angeles. I flew down from San Francisco for an interview, and was hired that day. There was only one problem—I still had four months left on my contract with the Navy. This was no concern for Western, and they told me to let them know when I was ready for a class. In some ways, this was both good and bad. It was good in the sense that I still had time to decide if I really wanted to leave the military. I truly enjoyed the time I gave in service to my country, and I still had not decided if I would leave after completion of my five-year pact. It was bad in the sense that it gave me too much time to think.

After I obtained a job with Western, I began to wonder if there was something out there that was just a little better. I started to consider what the different airlines were paying. In conducting my research, I found out that World Airways in Oakland, California, was paying almost three times the starting salary of the others. This was the first time that I had ever considered the amount of money I would be paid for doing a job that I would have done for free. Even though there were things I did not like about their type of flying, I ignored the warnings from inside and applied for employment. Before I could be considered for the position, I had to undergo extensive written and psychological testing. On the morning the exams were scheduled, I once again became ill. When I arose in the morning, I considered postponing the tests until a later date. This is one decision I am glad I never made because I did not get hired. I was feeling so poorly that I couldn't concentrate on the written or psychological tests. A few days after this ordeal, I received a call that I didn't perform well enough to meet their minimum standards. This was another blow for my ego, but I survived as it was not the right job for me. The trips would take me away from home two weeks or more at a time. The flights were long distance flights, and I would not have enjoyed that because it reduced the number of landings I could make. This was the fun part of flying for me. At the time, I wasn't keeping score of all the times I tried to make things happen my way, but I was ending up in the losing column. Once again I stumbled and God was there to put me back on my feet. Many years later World Air Lines filed for bankruptcy protection.

During the four months until my release from the military, I looked at many other options. I was attracted by the worldwide reputation of Pan American World Airways, Trans World Airlines, and a few others. History has recorded the fact that the glossy picture of these airlines faded, and they are now history. They no longer exist. In my quest of finding the "right" job, I would often run into an obstacle which caused me to abandon my search. I resigned myself to the fact that I would be working for a "less than perfect" company. This turned out to be a great misconception because Western was a wonderful company for which to work. If I had been hired by any of the other companies, I would never have met my wife. This, in itself, has been the main reason that I have never had any regrets about things turning out differently.

My stint in the Navy ended in July of 1967, and I was in class for Western on August 21, 1967. While I was in training, I was also flying C-118s out of the Los Alamitos Naval Air Station for the Navy Reserves. This airplane was the military version of the DC-6. On one of my flying days in the reserves, I had the opportunity to meet one of the vice presidents for Air California. He had come to the station to find pilots to fly the new Boeing 737 they were to acquire. His mission was to obtain names of prospective employees from the commanding officer of the reserve unit. My name was one that was recommended, and the man approached me with a job offer. This was one of the most attractive proposals that I had ever encountered, and I turned it down. Afterwards, I wanted to pinch myself to see if I really was alive. The offer that he made was that I would be hired as a captain without ever having to fly in the positions of a flight engineer or copilot. At the time, I was in training for the flight engineer program at Western, and I knew that it would be a few years duration before I moved into a copilot slot. I occupied the flight engineer seat for over four years, and it took a total of eleven years to have enough seniority to check out as captain. The reason that I gave for turning down the offer was because of the schedules for Air California. They only flew in the state of California at the time. My desire was to expand my horizons.

About a week later, I received a telegram from American Air Lines offering me a job in their next class. American had a much better reputation, and this was another enticing offer. I simply told them I was happy at Western and I would be staying there. With my history of trying to find something that was just a step above the rest, here I was turning down what I would have gladly accepted at an earlier time. God was beginning to refine my decision-making process. I don't understand why I was at peace with each of the last two choices, but I focused my attention towards the job I had with the intent to give it my very best. Many years later Western merged with Delta, and this gave me the blessed opportunity to work for two great companies.

> Not that I have already attained, or am already perfected; but I press on, that I may lay hold of that for which Christ Jesus has also laid hold of me.
>
> Philippians 3:12

Even when I didn't have any idea of God working in my life, He was working behind the scenes. This highest form of being faultless is only achieved when we go to be with the Lord. When I accepted Christ as my Lord and Savior, I became perfect in the sense of my standing with Him, but I still have an enormous amount of maturing to accomplish before I one day fly my last flight and enter into that state of ultimate perfection.

Am I still content with the way things worked out in my life? The answer is a resounding "yes." I recently played a round of golf with a college professor who specialized in sport psychology. He is on retainer for professional sports teams, and he works with the U.S. Olympic athletes. I asked him what he told his clients after they had a bad game. He said, "I tell them that they can't live in the past." Someone once said, "Yesterday is history, tomorrow is a mystery, and today is a gift. That's why it is called the present." If we look to God for answers to life's questions, He will provide.

My Testimony

> Peace I leave with you, My peace I give to you; not as the world gives do I give to you. Let not your heart be troubled, neither let it be afraid.
>
> John 14:27

Have you ever been asked the question, "Where were you when some dramatic thing took place in history?" Where were you, if you are old enough, when Japan attacked Pearl Harbor? Where were you when President John F. Kennedy was assassinated? Where were you when Martin Luther King, Jr. was gunned down at a motel in Memphis? Where were you when terror struck at the heart of America in New York City on September 11, 2001?

When John Kennedy was shot, I was flying overhead near Dallas on a training flight out of Corpus Christi, Texas, in a S2F training aircraft. Upon landing back at Corpus Christi Naval Air Station, the flag was flying at half mast. One of the guys with whom I flew remarked, "I wonder what happened while we were gone." Off the top of my head I said, "Maybe someone shot the President." We would find out in a few minutes after parking the airplane that this was indeed the tragic reality. When terror struck the twin towers of the World Trade Center in New York, I was home in Salt Lake City. I was checking the news on the Internet when I saw that a plane had hit one of the towers and smoke was billowing out the side of the building. I immediately went to the TV and watched in horror at what was to come in the next few minutes.

History records these events as significant tragedies in our nation's past. Why do I mention these things before I give you my testimony? There is something within us that helps us recall things that are important and full of meaning. While these images still reverberate in my mind, they are not the most important thing in my life worth recalling. They are interesting to discuss from time to time, but they are just events, horrible as they were. The most important event in my life was the night that I yielded my life to the Lord Jesus Christ. It was on a Wednesday night in late May of 1969. I thank God that I can recall every significant thing that took place that night. Let me share my testimony with you.

I grew up in a small farm community in Tennessee, seven miles from the town of Sneedville. My parents, Kyle and Juanita Lawson, owned a tiny farm of about fifty acres. My dad managed a hardware store in Sneedville. My mother took care of household chores, managed the farm and was the primary caregiver for three boys: Gerald, Danny, and myself. Twenty-three and one-half years after Danny was born, my parents had a daughter, Anita. We had a renter who lived on the farm, and he did most of the heavy chores connected with maintaining the farm which included the tobacco crop.

Every Sunday morning we would all go to church at Duck Creek Baptist Church. This was our weekly ritual. The unspoken rule in our home was that you didn't do anything on Sunday until you had gone to church. We occasionally went to Trents Chapel Baptist Church where my grandparents attended and where my dad went to church in his early years.

About once a year, someone would come to one of these little country churches to hold a revival meeting. I attended one of these meetings in the winter of 1951 at Trents Chapel Baptist Church. I was ten and one-half years old at the time. When you reached the age of ten you were thought to be at the age of accountability. If you had not accepted Jesus Christ as your Savior and you died after reaching this age, many believed you were going to Hell.

At these revival meetings, the scenario went like this: The pastor, or evangelist, would give his message and invite people to accept Christ. After the sermon, those already saved would usually come forward for a time of fellowship and prayer. All the unsaved people in the pews were left as targets for those individuals who had gone to the front. I was one of those people in the back waiting for an invitation to come and join the fellowship of the brethren. I thought I was tough, and I knew I could resist any invitation that was given to me. I hadn't counted on my grandmother, Bertha Lawson, as the one who would invite me to come forward. Although I didn't fully understand the meaning of going forward, I couldn't say no to my grandmother. I went forward and after about an hour of people praying for me, I got up and said that I had accepted Jesus as my Lord and Savior. I did this because they said they would stay there all night if they must in order for me to make this decision. By standing up that night, I began to live a lie that lasted for almost seventeen years.

It was not until I was thirteen years old that I realized that I was truly a lost sinner destined for Hell if I didn't repent of my sins and turn my life over to Christ. Because of what I had done three years earlier, I gave excuses to God for the next fourteen years before I was driven to my knees. I call this my "fourteen year close call" because I came face-to-face with danger more times than I prefer to remember during that time, but God somehow saw fit to exercise his patience with me. I faced life-threatening situations in Vietnam and other places as well. I don't understand it all, but I am glad God gave me more than my fair share of chances to accept Him.

During my "fourteen year close call," I continued to attend church when I could. I was constantly aware of the decision I needed to make to set things straight with the Lord. My excuse was always, "I'll do it tomorrow." I had a strong belief in God, Jesus Christ, and the Bible. At the age of ten, I had this same belief structure. I was not convinced at that age that my sins were so bad that they needed forgiveness. It was when I reached thirteen that I realized that I should have gone to the cross to accept Jesus instead of trying to run. From that time on, there was never a doubt in my mind what I needed to do, but I still put it off for fourteen years and allowed Satan to use many tactics and devices to keep me from giving in to the Lord. One device Satan used was convincing me that I would look like a fool if I admitted that I had deceived those people in church when I came forward. I must say this was a clever tool of Satan because I was heavily influenced by it.

In October of 1964, more than two years after I joined the Navy, I was married. During our marriage, my wife and I went to church once in a while. The longer we stayed married the less we attended church. In February of 1969, my wife went to see a lawyer and filed for divorce. She was three months pregnant with our son Kevin who was born six months after we separated. This was the first divorce in my family, and I was devastated. I didn't think I was bad enough for my wife to leave. I didn't realize all of my contributions to the divorce until after I accepted Christ. Although I was a good provider and a faithful husband in the worldly sense, I did not meet the emotional needs of my wife. The Bible is very clear on the subject of divorce: God hates it. From my perspective, I had a biblical reason for the separation. I didn't walk out on her, but I was a major contributor to her walking out on me. This was the hardest thing I had ever experienced in my life up to that time. One of the strongest pieces of evidence that proves God hates divorce is the fact that you live with the hurt for years to come. The ugly consequences of sin reside with you for the rest of your life. God forgives, but the scars remain to remind us what happened.

Shortly after my wife left, a neighbor invited me to attend church with him. I told him that the last thing I wanted to do was go to church. Upon turning him down, he continued to try to get me to come with him. He invited me to a prayer breakfast on Saturday morning, and I said that I would rather go to church than attend a prayer meeting. Even after this, he didn't give up. He said that they were going to eat at a restaurant after the prayer meeting, and he wanted to know if I would like to join them. I replied that I was going out to breakfast anyway so I agreed to meet them at the restaurant. At the breakfast, he introduced me to the pastor of his church whose name was Bob Willis. I enjoyed the time with his friends, and I liked the pastor. The pastor and I were at the restaurant until early afternoon. As we were getting ready to leave, he invited me to church the next day. I told him that I was working but that I would attend the following Sunday.

When that Sunday came, I went to church, and I felt like I had returned from a long vacation. I enjoyed the people, and they made me feel at home. The pastor took me under his wing and invited me to his home for lunch after church. He would do this often on the Sunday mornings when I attended. I had told him that I was not a believer but that I knew that I needed to one day accept Jesus Christ. My excuse was that I was just not ready yet. He continued to love me without putting pressure on me to believe, and this seemed to put me under more conviction.

Finally, in May of 1969 on a Wednesday night, I made the decision to accept Jesus Christ as my Lord and Savior. I distinctly remember, as the evangelist was playing the special music, "The New Jerusalem," I bowed my head and said something like this: "Lord, I have spent twenty seven years running my life my way, and it is not working very well. I want You to forgive my sins and come into my life and take control before things get worse. I believe that You are the Son of God, that You died for me on the cross and that God raised You again on the third day. I believe You are who the Bible says You are." After confessing this, there was a peace that came over me that I can't explain. When I went to church that night, I knew that if my life ended I would go straight to Hell. After making the decision, I knew that my destination was changed, and I would go to Heaven if I died. I turned in my old ticket and exchanged it for another that would take me to a much better place.

Let me tell you some of the things that changed when I took the step to become a child of the King. Before this time, I was continually looking for that one thing which would fill the void in my life. First, I thought everything would be perfect after I graduated from college. I did that and "something" was still missing. After entering the Navy, I thought I would achieve the happiness for which I searched by getting my wings as a Naval Aviator. Again, this did not fill that empty spot. I sought one goal after another because I knew that one of them was going to fix everything. Some of those goals were: flying my solo flight, landing on an aircraft carrier, getting my instrument rating, and finally graduating from flight training. I was gratified after reaching each of these stages, but they did not bring that extra measure of contentment that I so desperately wanted. After getting my wings, there were other goals that I thought would solve my problem such as C-130 pilot training, aircraft commander training, and on and on. None of these provided true satisfaction and happiness.

The answer came on that Wednesday night in 1969 when I invited Christ to come into my life. No, this did not solve all the problems I was having at that time, but the peace did come. This meant that I could end the elusive search for that "peace and happiness." With Christ in my life, I now possessed that incredible gift. The same is available to everyone who will just repent and believe.

My Fourteen Year Close Call

A question about flying I often receive is, "Have you ever had any close calls?" Most people, who have lived for a significant period of time, have had one or more "close calls." When it comes to aviation, there seems to be a heightened desire to hear about a brush with death. I don't know why this interest exists, but there could be many reasons. One may be that people want to hear that you have never had a close call, or they just want to be reassured that it is safe to fly. It is not my intended purpose to make this a psychological evaluation of why people ask this question. I will let that remain a mystery. I want to tell you about some of my close calls during a fourteen year period of time, from the age of thirteen to twenty-seven, when I was running from God. This is why I call this "***My Fourteen Year Close Call.***" During these fourteen years, I was not a Christian. Why God allowed me to survive some of these things is somewhat of a mystery to me, but I praise His Name for being patient with me until I reached the age of twenty-seven. This is when I chose to quit trying to do everything on my own and turn my life over to Him. I simply said that I was sorry for trying to run my life on my own, and that I wanted Christ to come into my life. I wanted to start depending on the One who could make a real difference. God made many attempts to get my attention, but I was too busy with all my excuses.

I am reminded of what an American Airlines pilot told me several years ago while riding in my jump seat from Washington, D.C., to Atlanta, Georgia. We were discussing some exciting moments about raising children, along with some of the unforgettable things they often say. He was relating a story about a recent time when his four year old daughter had been misbehaving. He told his precious child that he was sorry but that her actions had necessitated a time out. Upon hearing this dreaded statement, his daughter replied, "Daddy, I don't think I can find time in my schedule for a time out." When God is trying to get our attention, we sometimes use the similar excuse, "I just don't have time to consider that now." When it comes to listening to God, there is no better time than the present.

> For He says: "In an acceptable time I have heard you, and in the day of salvation I have helped you." Behold, now is the accepted time; behold, now is the day of salvation.
>
> 2 Corinthians 6:2

Several years ago, I had the opportunity to be a discussion leader in the Bible study called *Experiencing God,* by Henry Blackaby. In that study I was reminded again that God speaks to us in a number of ways. He speaks to us through His Word, His Spirit, prayer, other people, and circumstances. I think you will clearly see that God was trying to speak to me in the following accounts. All I had to do was listen and respond.

One such close call was north of Pensacola, Florida, in 1963. I was practicing touch-and-go landings where you land and takeoff without coming to a stop on the runway. After making several touch-and-go approaches and landings, I was in the pattern to do another. When entering this pattern with other airplanes, you must adhere to strict procedures to insure the safety of all involved. If anyone chose to enter the pattern in violation of the established procedures, he could put himself and others in great danger. Another pilot decided to do this, and it almost cost me my life.

In the early and mid-1960s, the United States trained many South Vietnamese pilots for the war. Many of these people had very limited knowledge of the English language, which sometimes created problems due to a lack of understanding of all the rules. I was in the landing pattern when a Vietnamese pilot entered the pattern from the wrong direction. I was a new pilot concentrating on executing my procedures correctly and did not see everything outside the cockpit that I should have seen. I was flying solo, which meant I was the only one in my plane. When we were in the early stages of flight training, we would fly with an instructor one day to learn new techniques. The next day we would fly solo and practice those techniques. On this particular day, my regular instructor was flying just behind me with a student, and he observed what was happening.

I was flying parallel with the runway going downwind which means that I was flying in the opposite direction from where I would land. I was preparing to make my turn toward the runway when I noticed an aircraft in front of me that had entered the pattern at a point that was critically wrong. My instructor behind me actually put his hands over his eyes because he knew there was going to be a midair collision. We were flying so close to each other that a warning by this instructor would have been too late. I praise God that I saw the airplane in time, and I was able to execute an immediate climb to avoid him. We were rapidly closing in on each other, and the other pilot descended to avoid hitting me. If we had climbed or descended together, this would have ended in disaster. After I gained my composure, I wanted to find out who had violated this airspace. The other pilot realized he had made a mistake, and he tried his best to depart the area without anyone getting the number on his plane. I was determined to find out who he was so I pushed the throttle to the full position and chased the violator. After a few moments I spotted him and took the number on the wings. Upon landing at Pensacola, I went to the operations center to find out who was flying that particular craft and learned his name. I went to the individual with a stern reminder to never enter a landing pattern in that manner. He quickly acknowledged his mistake, in very broken English, and the matter was concluded at that moment.

God was speaking to me through these circumstances. A friend of mine was not so fortunate a few weeks later. He was in a similar landing pattern at Whiting Field outside of Pensacola. Another South Vietnamese pilot entered the pattern incorrectly just behind him. This other pilot flew his airplane into my friend, chewing off most of the tail of his craft with his propeller, and causing him to crash. This resulted in the death of my friend, though the Vietnamese pilot was able to maneuver his crippled plane to the runway without being hurt. My business is a lot like golf, “If you are going to play this game, you have to know the rules.” I have heard this phrase used by golf analysts on TV.

Let me digress for a moment and tell you about a couple of other situations that didn’t turn out so well. A friend came to me seeking advice because he had lost interest in flying, and had become fearful of it. I told him to go to his superiors and request a break from training so that he would have an opportunity to evaluate his circumstances before quitting the program. In a few hours after speaking to me, he was to go out on a solo cross-country night flight around the Pensacola area, which I advised him not to do. He told me that he was going to give himself a few more days to make up his mind. When it came time for the night flight, he was there. When the T-28s started to return from this mission, my friend was not among them. They found the wreckage of his plane the next morning with him still strapped in the seat. God is continually issuing warnings to us, and we need to heed them when He does.

Sometime later, I was having breakfast with a friend who was having second thoughts about his love for flying. Later that morning, he was to fly out to the U.S.S.

Lexington for carrier qualifications in an A-1 Skyraider airplane, a very high-powered, single engine craft. A couple of hours later, he was in the landing pattern at the carrier. As he turned toward the carrier, the landing signal officer on board the ship relayed to my friend that he was too slow. This meant that he was going to have to add power at the same time he was turning, in order to regain his speed. The job of the landing signal officer was to make certain that everything was just right before the pilot was allowed to land. As my friend increased the power, the torque produced by the large engine caused the airplane to roll. At this point on the approach, the pilot was only a couple of hundred feet above the water. When he started to roll, he was so low that there was insufficient altitude to recover. He crashed in the Gulf of Mexico just south of Corpus Christi, Texas. He did not listen to the small inner voice that nudged him to take the time to reevaluate his priorities. God speaks to us for a reason, but it is up to us take the necessary steps to resolve the issues.

From 1965 to March of 1967, I flew thirty missions to Vietnam from my home base at Moffett Field in Mountain View, California. During those missions, there were a few instances that could have had a much different outcome. On numerous occasions, I encountered unfriendly fire going into different bases in Vietnam. At night, it would sometimes appear that we were flying into a valley of rods. You could see the tracer bullets surrounding us, and you never knew if one of those bullets was going to find its intended target. When we landed during these times, we would sometimes pinch ourselves to see if we were really all right. One of the C-130E Hercules airplanes from our squadron, VR-22, was not so fortunate. The plane was hit and exploded in midair, killing all twelve crewmembers on board. We always lived with the anticipation that this could have been us, and I didn't understand why God permitted this to happen to them instead of me. Even though I was not a Christian at this time, I was aware of God and my deep need to have Him involved in my life.

There was another instance when my crew and I were flying into Saigon. We were being vectored, or guided, by approach control for our upcoming landing. Because of all the activity in the area, we had to make sure that we closely followed the instructions given to us. On this particular flight, I had two copilots. The nature of our mission dictated the need for one or two other pilots. One of them stood behind me in the cockpit filming our approach as our plane descended. At about six thousand feet above the ground, it was apparent that we were going to enter a cloud layer. Just before entering the clouds, the controller gave us clearance to turn left to a heading of two hundred seventy degrees which was a thirty degree left turn. After completing the turn, I told my copilot to report that we were now on the assigned heading. The controller abruptly responded by saying that he had told us to turn right to the heading of two hundred seventy degrees, as his intention was to give us a three hundred thirty degree right turn instead of the short, thirty degree left turn. My flight crew later confirmed that we had heard him correctly to turn left when he meant right. The cloud layer was only one thousand feet thick, and as we were getting ready to come back into the clear, the controller informed us that we were entering the middle of an air strike against enemy positions near Saigon. When we came into the clear, we were in the middle of about five jets dropping bombs, which my copilot captured on film. Once again, God cleared enough of a path for us get through without a scratch. This was another close call, and another chance for God to draw me to Himself. I still had more lessons to learn.

During my three years of flying the C-130E Hercules, I had to shut down engines approximately twenty-five times. If you were to talk to other pilots who flew this airplane, you would be hard-pressed to find someone who had experienced that many engine failures. God gave me plenty of warnings. Shutting down an engine on a four engine plane is usually no big deal. However, there were times when it was a concern. One of those times was on a flight from Wake Island to Okinawa. We were carrying about twenty thousand pounds of high explosive bombs to be used by our troops in Vietnam. The handling instructions on this cargo were very specific. If a fire broke out in the vicinity of this cargo, you were to remove yourself to a position several hundred feet away within a few seconds. When flying at a high altitude, it is difficult to comply with these instructions. The only way to accomplish this was to ditch the airplane in the water and swim as far away as possible. About two hours from landing in Okinawa, the fire alarm went off on one of our engines. This is usually of no great concern because we have fire extinguishers that can be routed to the engine to put out the fire. We went through the procedures for an engine fire and activated one of the two extinguishers. After actuating the switch, the procedure called for waiting a number of seconds to see if the fire was going out. We waited the required time, but the fire alarm was still present. At this time we fired the second bottle and waited: our options were running out. The fire warning remained, and we were down to a couple of choices. If the fire got worse, we would have no choice but to ditch the airplane in the ocean and hope to be rescued. I notified the U.S. Air Force at Okinawa, and they dispatched a search plane to come out and escort us the rest of the way to the base. This search plane had the capability of landing on water if the need arose. The fire did not get worse, and we met the rescue plane and flew on to a safe landing. This is definitely another time where the outcome could have been much different. I still had lessons to learn because I did not give adequate thanks to God for bringing me through another ordeal.

In the later part of the 1960s, I was on a trip to visit my family in Tennessee. On that visit, I had an opportunity to fly a single engine airplane with a friend of mine in my hometown. He owned a farm along the Clinch River near the town of Sneedville, and had an airstrip in a field between the river and the highway. This airstrip was essentially a farmer's field where he had cut down enough hay to land his airplane. This is the only place I have landed where the runway has a slight dogleg, and that made things a little interesting. There were also electric power lines to avoid near one end of the runway.

My friend told me that he wanted to take me to a place above town to show me a piece of land he was thinking of buying, which he thought might be a better place to keep his plane, a Cessna 182. The weather was absolutely beautiful on that day, and I enjoyed the stunning scenery. When in a small craft such as this, the landscape is much better seen than at higher altitudes. As we soared along, I took photographs out the side of the window. I remember thinking that we were flying awfully low as we proceeded down this valley. In a single engine airplane, you want to make sure that you are continually aware of a place to land in case the engine fails. This was no problem because we were alongside a road where we could have landed if necessary. I was busy taking photos at the same time as enjoying the view when, suddenly, my friend exclaimed, "Oh, shoot." He pulled back quickly on the yoke of the plane to begin an immediate climb. I realized he had seen a high tension electric wire that was suspended across the valley where we were flying. At the instant he began the climb, we hit the wire, and the propeller severed the line. The wire contacted the plane in the center of the propeller dome. The severing of the line inflicted slight damage to the wingtips as the wire snapped. If this had not been the case, we would have been involved in a serious accident with possible fatal consequences. The high tension wire would have caused the airplane to flip if we had hit it with the landing gear, and there was a real possibility that it could have sliced through the cockpit if we had been a little lower.

We later learned that my friend didn't see the power line earlier because the power company was behind schedule to cut down trees in the path of the towers supporting the lines, which made it difficult to see what was in our corridor. The warning for me was realizing how low we were flying, but I ignored it because I was focused on the beauty around us. This was an invaluable lesson for me over the years because it helped me become more attuned to warning signs that might be present. The only explanation about why I survived some of these close calls is because I believe that God had a plan for my life of which I was not aware. In an incident such as this, few live to recall the circumstances. God demonstrated His patience with me one more time.

There is one more story I will tell you. Early in 1966, we were returning from another mission to Vietnam. On this particular flight, we were going from Midway Island in the Pacific to Moffett Field in California. Along this route there is a point that is the longest distance from land than any other place in the world. We were near that point when we began to have engine problems. As I said before, losing one engine is not a great cause for alarm on a four engine aircraft. In the evaluation period of trying to determine a course of action regarding one engine problem, we began to have trouble with another engine on the same wing. On this flight, we had about fifty or sixty soldiers returning home from their year of duty in the war zone.

We now had to determine the seriousness of both of these problems. Our conclusion was that we could keep one of these engines running at a reduced power setting, but we were not so fortunate with the other. We had to shut it down. In doing so, this would mean that we had to go a rather long distance on two good engines and another that was operating far below its potential. This situation required us to notify the proper authorities of our dilemma, and the U.S. Air Force dispatched a search and rescue plane to meet us and escort us back to the United States. There were times during this episode when we discussed a number of things. None of these discussions included how God was working in this particular situation. We talked about the sacrifice of these brave passengers who had devoted a year of their lives to assure liberty and freedom for all Americans. Would they survive the atrocities of Vietnam and then meet their fate on an airplane ride over the Pacific? We did not dwell on this for a long time, but it did enter our conversation. We safely returned to our base in California, and we had a part in returning these soldiers to their families. In retrospect, I look at this experience as another time when God was trying to get my attention.

In his painting "An Allegory of Prudence," the 16th century Venetian artist Titian portrayed Prudence as a man with three heads. One head was of a youth facing the future, another of a mature man eyeing the present, and the third was a wise old man gazing at the past. Over their heads Titian wrote a Latin phrase that means, "From the example of the past, the man of the present acts prudently so as not to imperil the future." When I talk to people, I try to get a similar message across to assist them in making decisions about their future. Take a look at your own examples of past mistakes, or mistakes of others, to help you act in such a way as to have a positive impact on your future. Remember, God will try to get your attention in a variety of ways, but you have to be open to hearing His voice.

Military Moments

Some of the experiences from my military days come from a slightly different perspective, because I didn't become a Christian until almost two years after fulfilling my obligation for the Navy. I was aware of the fact that I had a deep need to make things right with God.

After I finished my flight training, I had over three years of service remaining. During that time I was stationed in Norfolk, Virginia, for about a year and a half. The remaining time was spent flying out of Moffett Field in California. It was from this base that I flew my thirty missions to Vietnam. My trips out of Norfolk were primarily to Europe, Cuba, and Puerto Rico. There were so many things that transpired during those three years that it is impossible for me to recall more than a fraction of them. I will just tell you about a few of them.

In July of 1962, a little more than a month out of college, I found myself on the way to Pensacola, Florida, to begin my naval service. When I arrived I was unprepared for what was to come. At the airport, I met a number of men who were scheduled to be in my class. It was the class of 28-62, the twenty-eighth class to go through Pensacola that year, and they were training pilots at a rapid pace at that time. This was the early preparedness stage for the Vietnam War. Of the men I met on the way to the base, I was the smallest one there. I was five foot six inches tall and weighed one hundred twelve pounds. When I took my original physical, I weighed one hundred twenty pounds. The doctors gave me eight weeks to gain six or seven pounds if I wanted to remain in the program. I gained all the required weight in the first week.

Shortly after arriving at the Naval Air Station, we checked into our barracks which would be our home for the next sixteen weeks. While I selected my bed, I heard the strong voice of the gunnery sergeant as he was walking directly towards me. He tapped me on the shoulder and said, “Green shirt, I’m going to see to it personally that you never make it through this program. If you do, I promise I will do back flips for you.” He called me this because I was wearing a light green shirt. Those comments would reverberate in my mind for the next few weeks. I must say that he wasn’t the only one who didn’t think I would make it through the program. There were many people who had told me that this program was just too tough for someone of my stature. This would prove to be a great incentive for me. When the going got rough, I would play the tape in my mind, “You’re just not going to make it.” This always gave me a little bit of extra determination each time I thought about it.

Even though there were some anxious moments during preflight training, I did make it through. The most difficult thing for me was passing the Class A swimming program. When I arrived, I didn’t know how to swim. It was not until the twelfth week that I satisfied all the requirements for this part of my training, and I squeaked through just before the deadline. Upon passing this test, the instructor said that he had never seen anyone try so hard.

Two-thirds of the way through preflight training, I had won the respect of the gruff drill sergeant. Some of the training was a little easier after that. When I received my commission as an Ensign in the Navy, the sergeant saluted me first out of the fifty people in my class. I was now ready to tackle the real flight training program. Until this time, we had still not been in any aircraft. This was just the rigid physical and military training.

The label of being the smallest guy around would not end with my preflight training. About a month later, I purchased my first car. It was a Dodge Polara 500 convertible. One of the first times I drove it through the city of Pensacola, with my top down, I was stopped by the city police. After he pulled me over, he asked for my driver's license. I knew I had not done anything wrong, and I was not speeding. I have always been pretty much of a speed limit driver.

Well, let me get back to the policeman who stopped me in Pensacola. He said the driver's license wasn't good enough. Bear in mind that I was in my officer's uniform while I was driving. He wanted to see my car registration. When he examined it, he started laughing. He said he didn't think I was old enough to be driving. At that time, I didn't appreciate his humor.

This wouldn't be the last time something like this would happen to me. A few years later, I was flying missions to Vietnam. I had also been promoted to aircraft commander, which in civilian terms is equivalent to an airline captain. It was 1966, and I was in Okinawa getting ready to take a load of cargo down to Vietnam and back. This was a very sensitive time in the war, and we were required to get top-secret briefings before entering that war-torn country. This was for our safety and for providing us with information that would be useful to us on the trip.

The briefings were given by a U.S. Air Force major. He was stationed in a little room that had a very small opening from which he dispensed his information. I walked up to that little window and said that I was there to get the briefing for Lieutenant Lawson's crew. This little opening was rather high, and I almost had to stand on my tip toes to see him. The major came to the window and announced that he needed to see Lieutenant Lawson personally. He was a rather large man, weighing more than two hundred fifty pounds. I informed him that I was Lieutenant Lawson, and he turned his head and began laughing. I told him I was used to a response such as this. He then composed himself and gave me the briefing.

The following incident happened when I was completing my training for aircraft commander. I had finished all the requirements successfully, but had yet to fly a typical mission from California to Asia with a regular crew. During the mission, I had a flight examiner with me for 7 days to evaluate my ability to do the job. I wasn't told until returning to my home base that I had passed the examination.

We returned to Moffett Field where we sat down for my debriefing to go over the last few days of flying. All I wanted to hear from the examiner was that I had passed. He began his assessment in a very serious tone. "Archie," he said, "It was a pretty good trip, and I have no negative things to brief you on. However, I am going to pass you under one condition." I was quite puzzled at that remark, and I said, "What do you mean?" He said, "You have to promise me to stay in the cockpit if you ever carry passengers. You will scare them to death if they see a little kid up there flying their airplane." With that I began to breathe a little easier. I guess I did look pretty young for my age. Oh what I would give to hear some of those comments now that I am somewhat more advanced in age.

As I said earlier, I was still not a Christian, but I had a strong sense of right and wrong. I didn't always make the right choice, but I knew what was right. This has always been one of the reasons that I have been a person who tried to live by the rules, and we had rules in the military. Even though I have never been someone who liked alcoholic beverages, I did have a drink once in a while with the guys. When we were flying, we had a hard and fast rule of no drinking within twelve hours of flying a mission. I never violated that rule, but that is not the case with everyone. Vietnam was a dark time in American history. There were many who used alcohol to help them get through. Here is one such story.

I was at the officer's quarters in Okinawa, and I had about two hours before I had to report to the base for another flight to the war zone. On this particular flight, I had two copilots because we had to put in some long days without crew rest. I walked into the room of one of my crew members, and he was drinking a beer. Near his bed, there were about four or five empty beer bottles he had already consumed. I did not say anything to him. When we were on the flights, I rotated the landings so that I would make one out of three landings. When it was this crew member's time to make the takeoff and landing, he climbed into the seat to assume his duties. I politely asked him to ride in the back of the plane and give his seat to the other copilot. This was all that had to be said, and he never gave me any more problems. He didn't have to be reminded about why he was being removed.

When we have done something that is not pleasing to the Lord, all He has to do is get across to us that we are out of fellowship with Him. At that time, we must decide to get back in step with Him or continue to live outside His will. God loves us, and He always welcomes us back when we repent and ask for His loving forgiveness.

> If we confess our sins, He is faithful and just to forgive us our sins and to cleanse us from all unrighteousness.
>
> 1 John 1:9

Boeing 720B

This was my first jet aircraft. I began my commercial aviation career in August of 1967 as a flight engineer and served in that capacity for over four years. Western Airlines introduced a fleet of thirteen Boeing 720Bs in 1961. This was the paint scheme at the time I began my profession. They would remain in service until the 1970s. The 720 got this designation from the 707 when it was modified for short to medium routes and for use on shorter runways. Engineers shortened the fuselage length by nine feet, changed the leading edge flaps, and later installed turbofan engines. Boeing built one hundred fifty-four 720s between 1959 and 1967. I was also qualified on the 707 which had its first flight on December 20, 1957.

Playing the Game of Flying

Before flying one of the modern day marvels, many hours must be spent in a flight simulator. A flight simulator is a series of computers working together to generate the re-creation of the flying experience. Sophisticated flight simulators, costing well over a million dollars, provide pilot training without putting human crews and planes at risk. They have the ability to simulate normal and emergency situations. The equipment responds like a real aircraft. This allows pilots to experience lifelike flight situations without the danger or cost of real flights. These marvelous playthings are so advanced that a pilot can accomplish all of the required training without ever going on board the actual plane. After successfully completing training, a pilot is certified to carry passengers the first time he actually flies the real machine. This has not always been true. Flight simulator software, running on personal computers, is more advanced today than ever. It simulates flight in a less realistic fashion while providing entertainment and practice in reading navigation instruments. This software will not equip you to fly the plane, but it gives a wonderful introduction.

In 1962, when I started flying, the simulators were a mere shadow of the ones in operation today. In the Old Testament, we have many pictures of things which are revealed later in the New Testament, and those new things became far greater than anything before them. The first trainer to which I was exposed was a mock up of the cockpit of a Beechcraft T-34 Mentor, the airplane in which I began my flight training. This device familiarized a future pilot with the instrument layout in the plane. There was no motion, or any duplication of the sensation felt in the actual aircraft. Once a pilot knew the layout, he still had to accomplish all of the flight training in the T-34. In thirty-nine years, I saw these training devices evolve to the point where they are today. Today, almost every imaginable flight condition can be recreated without ever leaving the ground. The student must prove to the instructor, or examiner, that he can safely fly the airplane in every aspect of normal and emergency operations to be certified to fly. To certify on a new type of aircraft, over thirty hours of simulator instruction must be undergone. In addition to this, many hours in less complicated mock up trainers are required to master specific skills. A total of four to six weeks are spent in an intensive classroom and simulated environment. This is still practice and preparation. It is not the real deal.

In order to be a productive pilot, all of the necessary requirements must be completed. Attending a Bible study is much like a practice session in an aircraft training device, in that the Bible student is prepared to go into the world and apply what is learned in the lessons. Some people never take what they learn out of the classroom. If a pilot remained in the training environment, without going out and applying the skills he learned, he would be totally useless. We have a directive from God to go beyond reading and studying in order to be productive believers.

I have had people tell me that witnessing for Christ is not for them. They claim that they are too unskilled to effectively proclaim the Gospel. In the 1980s, I attended a series of meetings in Indianapolis. One of the speakers told the story about a mission trip for Gideons International in Mexico City. One of his assignments took him to a prison for women where he was scheduled to speak to two hundred fifty women about the Gospel of Christ and how this would change their lives. All of the inmates spoke Spanish, and the speaker only knew the English language. He obtained a female interpreter who was only sixteen years old. During the presentation, the young lady wrongly conveyed part of the message. This was unintentional but all of the women broke out in laughter. The young helper sat down and she refused to continue. One of the people who accompanied the speaker was formerly a man of the street who had recently accepted Christ into his life. He was not equipped to teach an extensive Bible study, but could simply share how Christ had changed his life. About half of the women in the audience gave their lives to Christ when he finished. This man did not shy away from this opportunity by saying he was not good enough. He made himself available. This is what God wants from us.

When I underwent aircraft commander training in the C-130, the instructor had difficulty demonstrating one of the required maneuvers. He attempted several short field landings but was never satisfied with the results. I listened to his presentation of how to accomplish this and said, "I believe I have the idea down. Just relax and let me give it a try." I only needed one approach to perform the technique correctly. He then said, "I believe you have the right idea." I was not a better pilot than my instructor, but God can use his Word in ways we can never imagine, even when we think we are ineffective. We may conclude that we may not always give an A-1 presentation, but God will use our witness for his glory. The instructor was a highly accomplished pilot who was just having a bad day.

There was another time when I received instruction about how to join another plane in formation flying. The execution of a turn toward the other aircraft had to be precise. The instructor tried to demonstrate the technique, but had a tough time. He was having an awful "fly" day. Even though he failed to perform the maneuver properly, he explained it clearly. When I took control of the plane, I was able to join the formation correctly. There is no excuse to not witness for Christ, even if we fumble. We must allow people to see that He has changed our lives. If we are unable to tell others about this change, we must reevaluate our own salvation.

When it comes to serving Christ, there are two kinds of Christians. First, the seasonal believer is someone who is informal, careless, laid-back, unconcerned, occasional, and indifferent. You might call this Christian a casual worshipper. Second is the devoted, dedicated, loyal, faithful, and unswerving servant of God as illustrated by my flight training experience.

In 1990, I was in Atlanta training to fly the Boeing 757 and 767 aircraft. I was one of two captains in the class in addition to two copilots. Until this time, I had flown the Boeing 727, but compared to the 757 and 767, this was old technology. The first 727's were introduced in the mid-1960s. This transition from one type of airplane to another was the toughest that I had experienced, and I had to concentrate on my studies more than ever. In addition to the classroom work, I devoted more than three hours to homework each night. The other captain in the class assumed he could get by with attending the class and doing no homework, and it soon became apparent that he was lagging greatly behind the eight ball. Needless to say, he did not pass the course, and had to extend his training. This pilot was an example of a casual learner while I tried to apply myself to the task at hand and was faithful to complete it.

> And whatever you do in word or deed, do all in the name of the Lord Jesus, giving thanks to God the Father through Him.
>
> Colossians 3:17

There used to be a statement in the United States Air Force Training Manual that said, "A superior pilot uses his superior judgment to avoid situations that require his superior skills." Skill and knowledge are necessary ingredients of a good pilot, but proper judgment, which leads to a correct decision, is essential. Good judgment is more difficult to learn than flying skills. This is why simulator training is vitally important. The ability to make good decisions must be demonstrated before a pilot is permitted to carry passengers. A look at general aviation accident reports reveal proof of this fact. Judgment and decision making flaws are common in a high percentage of general aviation accidents. Some of the other factors are failure to pay attention, becoming distracted, and lack of skills and knowledge. The small plane pilot, flying recreationally, is not subject to the same stringent requirements faced by commercial pilots. One of the reasons why commercial flying is much safer today is because problems are detected and corrected before an airman flies the airplane with a load of passengers. If an airline pilot makes a mistake, he has a backup in his flying partner to help him correct the deficiency. The general aviation pilot is often flying solo, and he does not have the luxury of a backup.

Passengers often make judgments about their pilot by the smoothness of the flight and the landing. Neither of these, the smooth landing nor smooth flight, has anything to do with an airman's ability. Even the very best pilot will have an occasional firm landing, and the person flying the plane has little control over the smoothness of the flight, except to change altitudes to find less turbulence. But even then there are times when a smoother altitude cannot be found. This is not the fault of the person flying the aircraft. True ability can only be properly assessed by looking at all the areas of the flight. This includes the handling and operation of the aircraft, working effectively with the crew and passengers, and a commitment to safely transport the passengers to their destination.

Although I underwent extensive training in classrooms and simulators, I received valuable instruction in the cockpit working with highly skilled aviators and closely observing their performance. This is called "on the job training." Early in my career, I decided that I wanted to do the very best job of which I was capable. I was not born with a silver airplane in my mouth, and I knew early on that I did not have the natural abilities which some of my coworkers possessed. I compare this to my college experience where some of my fellow students were able to glide through without a tremendous amount of effort, but it was much more difficult for me. It took all of my effort to get by with a C+ average. In flying, I did not want to settle for a C+ career, and I vowed to give it my best. As I reflect back on my efforts, I am pleased with the results. God gave me everything I needed to fulfill my dreams and expectations in my job.

There is a tendency to want to skip over difficult parts of life and proceed straight to the fun. If I told you that I enjoyed all of the hours I spent flying simulators, I would be less than honest. I fully understood their purpose, and I knew this necessary part of my job would enhance my skills as a pilot. Years ago, our son Peter attended his first soccer camp. He was born with an incredible amount of natural athletic ability, but he came home very disappointed after the first day. Jody and I wanted to know why he was so discouraged. One of the coaches had demonstrated the trick of balancing a ball on the foot and knee. Peter wrongly believed that he could never be a good soccer player because he couldn't do this little maneuver. We sat down with him and explained he wouldn't become good at soccer by wishing, but by learning the rules and committing to a lot of practice. He began to practice, and in a few short hours had mastered the trick with the soccer ball. I have heard people say, "I wish I knew the Bible as well as that person." A richer, deeper understanding of the Scriptures will not happen without effort. The most brilliant truths are found by digging below the surface. Some marvelous things can be gathered when walking along the surface, but the better results are found deep.

"All work and no play makes Jack a dull boy" is an expression that has been around for a long time, but the Christian must consider moving from playing the Christian game to living a life that brings honor and glory to God. Attending Sunday school is a wonderful way to get new insights into God's Word, but the Word must be applied and practiced. If what is learned is never taken outside the classroom, then this is like the pilot who never takes his skills beyond the simulator. Truly regenerated people with a new life in Christ must takeoff and share this change with others. It's time to get your walk off the ground, and take it to new heights.

Douglas C-118

I flew this type of aircraft for almost two years in the Navy Reserves. Douglas developed this design for postwar airline service as the DC-6, making its first flight with American Airlines in June 1946. Civilian DC-6s were built along with 166 aircraft for the U.S Air Force and Navy. They were designated the C-118 and R6D Liftmaster respectively. One C-118 was designed with a VIP interior for President Truman and designated the VC-118 *The Independence*. Other VC-118s also served as Air Force One. I'll never forget the thrilling sound of those four radial engines as the plane gracefully raced down the runway and leaped into the sky.

Wartime Memories

The memories I carry from the war in Vietnam are nothing compared to the stories that countless others have to share. The war in Iraq has caused me to pause and reflect on my times of service. When I hear the news of some dedicated young soldier being injured or killed, I am reminded of the horrors I encountered. One of the types of cargo I flew into Southeast Asia was empty coffins. The thought of having to fly them out with human remains inside was very difficult. During the twenty hours of flying back to the United States, I had time to reflect on what was in store for the families of these young men. The war would not be over for these families for years to come because they would carry the sorrow and loss for the rest of their lives.

When a serviceman had completed one year of service, he could look forward to going home and returning to a life free from conflict and danger. Many of the soldiers who returned safely would carry memories with them that changed their lives forever. A large number today still live with physical and emotional problems they suffer because of the exposure to the war. I carried many of these people back from the battlefield.

I vividly recall the story of one Marine first lieutenant who was boarding the flight back to America. When I met this young man, he was lying beside the C-130. The sounds coming out of his mouth were unintelligible. I inquired about his condition to discern if I should summon medical help for him. His fellow serviceman told me that the young man was on the battlefield the day before with his best friend. This was one day before their one-year commitment was to end, and he watched in horror as his friend was struck by unfriendly fire. His friend died instantly in his arms. He had grown up with his best friend, and they did just about everything together. They had gone through elementary school to college, joined the Marines, were in officer's training and assigned to the same unit. As I look back on memories like this, I regret that I wasn't prepared to share the love of Christ with him, as I was in desperate need of His love at the same time. It would be several years later that I would accept the love He had to offer for my own life.

The streets of America have felt safe to walk until recently. There have always been isolated acts of violence, but there weren't moment-by-moment concerns about safety. While we remain the safest country in the world, there are increasing reminders that we are not as secure as we once were. In Vietnam, our world could be shattered instantly because of an enemy attack. My squadron flew into South Vietnam, which was much more secure than the North. A minimum of three people supervised the unloading of cargo from the plane. They watched for enemy infiltrators who might work incognito alongside our friends, and attempt to plant explosives on the airplane. An airplane in my squadron was blown up just after departure. We never learned whether the explosion was caused by planted explosives or if it came from ground attacks. I never discovered explosives on any of my flights.

The loadmaster is the person who supervises the loading of the plane to determine if the weight of the cargo is distributed correctly. An airplane will not fly if this is not accomplished, especially when carrying heavy loads. One of my loadmasters told me that he had found explosives on a previous flight. Once they departed the base in Southeast Asia, he noticed a rolled up newspaper in the back of an empty airplane. He picked up the paper, which seemed heavy, and discovered a beverage cup with a hand grenade inside. The pin of the grenade had been pulled. If the loadmaster had picked up the paper another way, the explosive device would have fallen out and the airplane would have been history. He had someone open the rear cargo door, and threw the device outside where it exploded. God had his hand on this crew, and sadly, they didn't know it. They called it the "luck of the draw." My wife says, "There is no such thing as luck; either it is God's blessings or His challenges."

My flight engineer told me about his brush with death when departing the war zone after dark. Shortly after takeoff, he picked up his flashlight rather than turning up the lights in the cockpit. He didn't feel the batteries rattle inside which caused him to be suspicious, and so he did not turn it on. He looked inside of the flashlight and found it half full of plastic explosives. If he had turned the switch on, it would have exploded causing irreparable damage to the cockpit with the result being death for all aboard. This is another example of God's blessing in the midst of a potential disaster.

When we flew to Southeast Asia, we typically arrived with a full load of cargo, and we departed with an empty plane. We arrived in Pleiku, a town in central Vietnam, in the Truong Son (Annam Highlands). Pleiku was at the heart of many grave actions of the Vietnam War, including a guerrilla attack on a U.S. Special Forces camp in 1965. This provoked the first sustained U.S. air raids of the war. The runway was composed of metal strips on top of sand and it was not very long. After unloading the cargo, we were ordered to take a load of critical supplies back to Okinawa. We did not have reliable data to determine if we could takeoff with that load on a short runway. We could only make a rough estimate of our capabilities and refuse to depart if we were positive the plane couldn't handle the load. There were many times we were asked to do things in a wartime situation that placed undue stress on the airframe. The C-130 is an incredible machine, but there is a limit to its abilities. We evaluated the runway length in relation to wind, the weight of the cargo, fuel load, and total takeoff weight, and arrived at the decision to honor their request.

We taxied to the runway, received clearance for takeoff, and began our roll. As we approached the end of the runway, it was apparent we needed more length, but were committed to takeoff. At the end of the runway, I rotated the airplane toward the sky and we became airborne. The liftoff appeared to begin well but we lost a fraction of our airspeed. This resulted in a loss of altitude, and the wheels gently touched the sand beyond the end of the runway. I could only keep pressure on the control wheel and hope and pray we would start flying. We barely touched the sand and the airplane jumped back into the sky and we were safely airborne. We breathed a sigh of relief in yet another exciting moment in the "glamorous" life of a pilot.

My average stay in Vietnam was around two hours, which was the time necessary to unload, refuel, and then return to Okinawa. The one time I spent four hours in Vietnam, I had the opportunity to see my brother, Danny, who was stationed in Saigon for one year. I was able to be reunited with him on only one of thirty missions, and could only later appreciate that it was the gracious hand of God who allowed for this time.

As I was making my approach into Tan Son Nhut Airport near Saigon, now Ho Chi Minh City, I flew erratically to avoid enemy fire from the ground. Upon landing, another airplane was still on the runway but was turning off at the far end. Under normal FAA regulations, I would not be permitted to land under these circumstances. The landing strip must be clear before permission is granted to land. In wartime, the rules are not so inflexible. After clearing the runway, I taxied the plane behind a Pan American Boeing 707 that had landed a short time before. I later learned that this aircraft had blown a tire upon landing. The only replacement tire had to be flown in from

Clark Air Force Base in the Philippines. The long lines of planes were forced to wait for four hours until the tire arrived. I give God praise for the four hour delay that allowed me the time with my brother.

My first trip to Southeast Asia was in 1965, precipitated by the guerilla attack on Pleiku. The route that I took on this trip was first to Hickam Air Force Base in Hawaii, then on to Wake Island, Guam, and Okinawa. When we landed at Anderson Air Force Base in Guam, the first B-52 strike on Vietnam was getting ready to depart. The mission would take them a total of twelve hours and included being refueled at least once before returning home. We watched the bombers depart for their mission and then took our required twelve hour rest time. Upon completing this rest, we saw the B-52s as they returned. We learned that two of the planes did not return due to a midair collision once the task of dropping the bombs was complete. A couple of people parachuted to safety after the collision but the remaining crew members perished, including an Air Force General. A few years later I learned details about that mission as one of the people who survived became a coworker at Western Air Lines. He told me that the crash happened because the airplane with the General on board had turned the wrong way.

A pleasant memory during my time in the military was running into one of my cousins in Guam. Clarence Green was stationed in Guam with his wife, Myrtie, and their children. Clarence is the son of my mother's uncle, Milford Greene. I had the opportunity to go to their house a couple of times when in Guam, and I played golf with him. This was a pleasant distraction from the normal trips I had to Vietnam. When I visit my family in Tennessee, I always look forward to spending time with my cousin and his family. He retired from the Air Force with over twenty years of service. I don't usually give him a lot of warning when I go back there because he has said that he would like to know when I am coming so he can make sure he is out of town.

All of the incidents included in this chapter were during the “fourteen year close call” that I wrote about earlier. I have only touched the surface of all the memorable military moments to which I was exposed during my five year military career.

Search for the Lost

> "What man of you, having a hundred sheep, if he loses one of them, does not leave the ninety-nine in the wilderness, and go after the one which is lost until he finds it?"
>
> Luke 15:4

I flew on a trip from Okinawa to Camrahn Bay, Vietnam, in late 1966. We received a call enroute from air traffic control that they had lost radio contact with an Air National Guard C-124 aircraft. I asked them for the last known position of the plane and they did not know if the plane had crashed, or if it had made it safely to land. A request was issued to me and my crew to assist in locating the airplane. I had about an hour and a half that I could devote to the search effort, before I would have to head toward the base and land, as we didn't have a large amount of extra fuel. The interesting little kicker to this was that the lost airplane had shut down one engine, and they were in the process of losing another at the time they lost radio contact. There were numerous reasons to be concerned about the safety of this mission.

We proceeded directly to the last known position that was given to us, and we did a square search for the hour and a half. From the position we were given, we kept expanding a square pattern outward to attempt to locate the lost aircraft. We were unable to find any trace of the plane and it was now time for us to abandon the search and proceed to our destination. It was difficult to leave without finding them, but we didn't have a choice because of our fuel situation. When we landed at the base, we were greeted with a pleasant surprise. The C-124 had landed safely with two out of four engines working. While in the air, they lost all of their radios which prevented contact with anyone on the ground. They proceeded to the base with a total loss of communication. We had an opportunity to meet the crew while we were in Vietnam. They were from an Air National Guard unit in Nashville, Tennessee. As I am from Tennessee, I almost felt like I was attending a homecoming reunion.

If I had been the one who was unaccounted for, someone would have done their best to search for me. Sometimes, many lives were placed in jeopardy when they undertook a mission to find a downed aircraft that was lost in the jungles of Vietnam. This compares to what the Lord will do if we are the one who is lost. I thank God that He sent out a search party for me when I was confused and disoriented. Unlike me, He didn't give up in an hour and a half. He never runs out of fuel. I knew I was lost, and I knew I wanted to be found. The first step in being found is admitting you are lost, and you must know that you cannot save yourself. After you admit you're lost, just believe that He has the power to rescue you.

Do you believe someone would come for you if you were the one who was lost? Consider this little story if you are placed in a position to rescue someone. I recently found this story titled *"It Was Worth It"* on the internet. The author is unknown.

Horror gripped the heart of the World War I soldier as he saw his lifelong friend fall in battle. Caught in a trench with continuous gunfire whizzing over his head, the soldier asked his lieutenant if he might go out into the "No Man's Land" between the trenches to bring his fallen comrade back.

"You can go," said the Lieutenant, "but I don't think it will be worth it. Your friend is probably dead and you may throw your own life away." The lieutenant's words didn't matter, and the soldier went anyway.

Miraculously he managed to reach his friend, hoist him onto his shoulder, and bring him back to their company's trench. As the two of them tumbled in together to the bottom of the trench, the officer checked the wounded soldier, and then looked kindly at his friend. "I told you it wouldn't be worth it," he said. "Your friend is dead, and you are mortally wounded."

"It was worth it, though, sir," the soldier said.

"How do you mean, worth it?" responded the lieutenant. "Your friend is dead!"

"Yes sir," the private answered. "But it was worth it because when I got to him, he was still alive, and I had the satisfaction of hearing him say, 'Jim, I knew you'd come.' "

The following story, "What is Really Important" illustrates the truth of taking time to consider others as God takes time to consider us. By Unknown Author from *A Cup of Chicken Soup for the Soul,* Copyright 1996 by Jack Canfield and Mark Victor Hansen, All rights reserved.

A few years ago at the Seattle Special Olympics, nine contestants, all physically or mentally disabled, assembled at the starting line for the one hundred yard dash. At the gun they all started out, not exactly in a dash, but with the relish to run the race to the finish and win.

All, that is, except one boy who stumbled on the asphalt, tumbled over a couple of times, and began to cry. The other eight heard the boy cry. They slowed down and paused. Then they all turned around and went back, every one of them. One girl with Down's syndrome bent down and kissed him and said, "This will make it better." Then all nine linked arms and walked together to the finish line.

Everyone in the stadium stood, and the cheering went on for ten minutes. Would you be the one to take the time to help someone, if you were faced with the opportunity to do so?

While I was in flight training at Pensacola, there was an incident where a pilot got lost on a solo flight. When he realized that he couldn't find his way home, he called the control tower. Instead of admitting his predicament, he requested a practice DF (direction finding) steer, a process where the pilot presses the radio transmitter button until the tower can home in on the location of the aircraft. After the tower locates the lost plane, directions are given to bring the lost pilot home. After the request was made, the tower informed the airman that he was too busy to conduct practice maneuvers. At this point, the student had to make a decision. He could continue to hide the fact that he was lost, or he could admit that he was in dire trouble. If he had chosen the former, he would have eventually run out of fuel and crashed. He chose the latter by telling the tower that he was in need of a "real" DF steer as he was lost and needed to find his way home. Without hesitation, the tower operator instructed the pilot to key his microphone so he could help bring him home. After doing this, his position was determined, and he was given the heading to fly to the airport. This is the way it is with someone who is lost and needs to find the One who can bring him safely home. The only requirement is to admit our lost condition and be aware that we cannot save ourselves. We must place our faith and trust in Jesus Christ, the ultimate Control Tower Operator.

Extraordinary Excuse

Over the years, I have heard excuses for just about everything. I heard an excuse in Billings, Montana, that I had never heard before. I was on a layover, and I went to dinner with the other two pilots and three flight attendants. Before I tell you about the excuse, let me provide the background information which leads to the discussion that took place at this dinner. On a previous trip to Billings, about a week earlier, I had flown with the same crew, pilots and flight attendants. Our trip took us from Denver to Cheyenne, Casper, Sheridan, and then Billings. On the stop in Sheridan, we picked up a passenger who was a sixteen year old girl on her way to see her sister in Billings. She had recently undergone the horrible ordeal of having been raped by her father.

The flight attendants noticed her dilemma, and took her under their wing. Upon arrival in Billings, the girl's sister was not at the airport, so the girl left a message at the airport and went to the hotel with the attendants. One of the flight attendants knew that I was a Christian and asked if I could come to one of their rooms and talk to the young girl. She said that they didn't know what to say to bring her comfort. When this incident happened, I had been a Christian about three years and didn't know how much help I could be. I did understand enough to know that God's Word could bring comfort when I couldn't.

At this early time in my Christian life, I did not carry my own Bible with me on my trips, but I knew that I could find a copy in the nightstand placed there by the Gideons. This fact was instrumental in my joining Gideons International in 1980. I spoke with the young lady for a while in the presence of the attendants, and then I took out the Bible and asked if I could read a few verses. These few verses, mixed with some discussion, lasted about two hours until the sister arrived to take the young girl home with her.

Before leaving, the young girl, her sister, and the flight attendants thanked me for spending this time with them. I cannot remember what I talked about that night. This will be one of the interesting things I have to look forward to in Heaven, because I will be reminded of those words sometime during the magnificent time in eternity.

I received several letters over the next few months from the girl, and she mentioned how much that time in the hotel room had meant to her. She said that she had experienced a peace during that time that she had never known.

Now, let's get back to the dinner in Billings. We had ordered dinner, and the waitress brought drinks. This was a rather diverse group of individuals. The captain was a Mormon, the flight engineer was an unbeliever, and I was a Christian. One of the flight attendants was an agnostic, another was a strong women's liberation advocate, and the other claimed to be an atheist.

Nothing had been said about religion by anyone at the table, but these ladies remembered what took place in the hotel room the previous week. After the drinks were served, the agnostic flight attendant slammed her fist on the table and made this shocking statement. She said, "I don't want to have anything to do with a God who would curse a fig tree." Everyone’s mouth dropped open. I responded, “I’ve heard a variety of reasons why someone didn’t want to have anything to do with God, but I have certainly not heard this astonishing excuse.” She was referring to the following verse:

> And seeing a fig tree by the road, He came to it and found nothing on it but leaves, and said to it, "Let no fruit grow on you ever again." And immediately the fig tree withered away.
>
> Matt. 21:19

People use excuses for a variety of reasons. They are used to release them from an obligation or responsibility, or to provide an explanation for their behavior that makes it appear more tolerable or less offensive. It can also be a phony reason that permits them to do what they want to do, or to avoid something they do not want to do. Consider the possible reasons for using excuses as I tell you this story, as it may lead you to examine yourself regarding times when you have offered excuses for the wrong reasons. What does God's Word have to say about the use of this term? We must carefully consider what God wants to tell us in His Word.

> But they all with one accord began to make excuses. The first said to him, 'I have bought a piece of ground, and I must go and see it. I ask you to have me excused.' And another said, 'I have bought five yoke of oxen, and I am going to test them. I ask you to have me excused.' Still another said, 'I have married a wife, and therefore I cannot come.' So that servant came and reported these things to his master. Then the master of the house, being angry, said to his servant, 'Go out quickly into the streets and lanes of the city, and bring in here the poor and the maimed and the lame and the blind.' And the servant said, 'Master, it is done as you commanded, and still there is room.' Then the master said

> to the servant, 'Go out into the highways and hedges, and compel them to come in, that my house may be filled. For I say to you that none of those men who were invited shall taste my supper.' "
>
> Luke 14:18-24

I told the young lady that I had only been a Christian for three short years, but I had taken an interest in reading the Bible. I said that I was not an expert, but if she would permit me, I would explain what I believed that short passage meant. She was agreeable to this, and nobody at the table objected to my explanation. In addition to my brief interpretation, I told her that I would do further research on this passage and put it in her mailbox the next time I came to work. I proceeded to do that research, and I placed in her mailbox a two page, single-spaced typewritten explanation, telling her my beliefs about this passage. When I gave this explanation, I had no idea of the impact this would eventually have. It was not until six months later that I realized that God had used this young lady's excuse for His glory.

> So shall My word be that goes forth from My mouth; It shall not return to Me void, But it shall accomplish what I please, And it shall prosper in the thing for which I sent it.
>
> Isaiah 55:11

This is a verse that has become meaningful to me over the years, and I am reminded of it each time I retell this story. About six months after the dinner in Billings, I was in Minneapolis, Minnesota, preparing for a flight in the lounge with several other pilots and flight attendants. A flight attendant approached me and asked if I was Archie Lawson. I said that I was, and she threw her arms around me in front of all the pilots and flight attendants. My face became beet red because I thought the people might think that I had some sort of relationship with this young lady. I told her that the gesture was nice, but that I was a little puzzled by her action. She proceeded to explain why she was glad to meet me, and asked if I recalled a trip to Billings about six months earlier. I told her that I had made several trips to Billings during that time frame, and I asked if she could refresh my memory about the specific trip. She mentioned the one where I had gone to dinner with the flight attendants and pilots and had heard the excuse about the fig tree. I told her that I remembered that trip very well.

This young lady told a story that has been an encouragement to me for more than thirty years. She said that the flight attendant who used the excuse of the fig tree was a very good friend of hers. In the course of a discussion about that evening in Billings, the agnostic flight attendant told her friend what I had said at that dinner. She also showed her the typewritten explanation of the verse in Matthew 21:19. This young lady said she became a Christian because of what I had said to the agnostic. She was a delightful young lady, and my heart was blessed to hear how her life had been turned from despair to joy because God's Word does not return empty. My friend, God, is faithful, and He will produce the fruit if we are willing to plant the seed and do a little watering.

There is a side note to this story. When I delivered the written explanation of the verse to the flight attendant's mailbox in the flight attendant's lounge, I was confronted by three of her associates. They saw me placing something in a mailbox, and assumed that I had some sort of relationship with one of the female cabin crew members. One of the young ladies said, "Oh, I see that you're delivering a love letter to a friend." Along with the letter of explanation of the verse, I also placed a copy of the Gideon Personal Worker's New Testament in her mailbox. In response to the "love letter" statement, I said that I was filling her box full of love letters, and I pulled out the New Testament. As I held God's Word in my hand, I said, "This little book is filled with God's love letters to man and it contains the greatest love story ever written." One or two of the three ladies were Christians, and they also requested a copy of this little Bible, to which I complied. God used the excuse of the agnostic flight attendant to get His Word out in quite an unusual way. You see, God will provide the opportunities if we are open and available to be used in a variety of situations and circumstances. This was beyond my wildest imagination. I praise Him for placing me in a position where I could bring glory to His wonderful name.

Memorable Moments

I am frequently asked if I have had famous people aboard my flights. Early in my career, I flew many times between Los Angeles and Las Vegas. There was hardly a flight that didn't have some "famous" person on board, and I often had notables on flights to many different destinations. Most of these people were very pleasant to serve, but quite a few made their presence known in negative ways. There were times when some of these people were self-absorbed and considered themselves too important to speak with "servants." I am referring to flight attendants. A famous person, her name known to most people, once informed a flight attendant that she didn't speak to the hired help. This puzzled me because she communicated with the flight attendant through her maid, her hired help, who was sitting beside her in first class.

I have many more positive memories than negative. I suppose the flight attendants could tell another version about some of the passengers because they were with them throughout the flight. The pilots were locked in the cockpit, and usually were called upon when things got out of hand. There were times when stern authority had to be exercised in order to control unruly patrons, whether famous or not. I was delighted to hear about how pleasant some people could be, and it is easier for me to recall the people who went out of their way to be nice, such as Bob Hope.

I vividly recall one of the most memorable famous people that I ever had the distinct pleasure to carry. Her name was Ethel Waters. She wasn't as well known as many of the others, but she had more of an impact on the people around her than anyone I can remember. She was a black actress who was also a great singer. She had been part of many Billy Graham Crusades, and she "walked and talked" her Christianity in a way that people could see and understand. She boarded the airplane in a wheel chair. I had the opportunity to observe her radiant character for a few moments, and was amazed at her interaction with those around her. People were drawn to something about her. It wasn't the way she was dressed. She did not wear a Hollywood designer outfit. This flight was one of the few times in my career when all of the flight attendants told me about the positive experience they had serving this passenger. Ethel Waters was a true living portrait of a disciple of Christ.

In the early 1970s, my crew and I flew from Los Angeles to Phoenix in the cabin as passengers to cover a trip out of Phoenix to Minneapolis. I was the flight engineer on this crew, which is the position in which I began my career. After obtaining enough seniority, I attained the position of first officer and then captain several years later. My two crewmembers and I boarded the flight before other passengers, and we were assigned seats in the first class section. The flight attendant had taken our uniform coats so that the other passengers did not know we were a "deadheading crew."

The first class section had just two other occupants: Governor Ronald Reagan and his body guard, a California Highway Patrolman. The copilot and I sat in the seats directly in front of them and the captain sat across the aisle. As they boarded the airplane we said a gentle hello, but we did not have any conversation with them until later.

Upon arrival in Phoenix, the flight attendant brought our uniform coats as the plane taxied to the gate. After Governor Reagan saw that we were pilots, he remarked, "Aren't you guys sitting a little far back to fly the airplane?" I replied by saying, "Governor, you would be amazed at the sophistication of the automatic systems on these new airplanes. We can fly them from back here." With that he had a big laugh. He had come to Phoenix to join his wife, Nancy, and her family for Thanksgiving. I had the opportunity on a number of occasions to carry him again, and always saw him with a smile on his face.

A few years later, Ronald Reagan became one of the most popular presidents in history. This story about him points out the need to make decisions without hesitation. This account contributed to how he became a great decision maker. An aunt had taken him to a shoemaker to have a pair of shoes made. The man asked young Ronald Reagan, "Do you want a square toe or a round toe?" Reagan hemmed and hawed. So the cobbler said, "Come back in a day or two and let me know what you want."

A few days later the shoemaker saw Reagan on the street and asked what he had decided about the shoes. "I still haven't made up my mind," the boy answered. "Very well," said the cobbler. When Reagan received the shoes, he was shocked to see that one shoe had a square toe and the other a round toe. "Looking at those shoes every day taught me a lesson," Reagan said years later. "If you don't make your own decisions, somebody else will make them for you!"

The sovereign God, King of the universe, has made us people with minds capable of making choices. He did not make us to be puppets. We have His Word to guide us, His love to redeem us, and His assurance that we are capable of making choices. We are not intended for running our life on autopilot. We need to turn the autopilot off from time to time and use the gifts God gave us.

Another memorable moment happened a few years ago when we prepared to leave on a flight from Salt Lake City, Utah, to Anchorage, Alaska. The cruise season had begun and flights were generally full during the summer months going to Alaska. People flew to Anchorage to board cruise ships going south to Vancouver, Seattle, or perhaps San Francisco. Some people chose to cruise north and fly back. We were oversold on this particular flight and the agents would have to entice several passengers to delay their trip for a few hours for compensation. Just before we were ready to leave, the agent came on board and asked three male passengers if they would be willing to give up their seats. They quickly and happily agreed. As they started off of the airplane, three women spoke up and said, "You can't take those guys off! They are our husbands, and we need them to carry our luggage."

This little incident started quite a commotion, and the ladies could not be quieted. I asked my head flight attendant to go outside and find out what was going on, and she returned with the reports of an interesting scenario. The three men had volunteered to fly at a later time so that they would not have to travel on the same flight with their wives. They had to be in Anchorage, however, within eight hours of the original arrival time to allow them to make their cruise. We left Salt Lake City with the three men still there. I can only imagine what the next few days of cruising had in store for these husbands. There is a good chance that a few unhappy moments awaited them there.

You have heard many stories of road rage, and you may have read about some instances of air rage. I never had to make an unscheduled landing to remove an unruly passenger, but I had to issue stern warnings to quite a few. In one such incident, I would have removed the offender had I known what was going on prior to departure. A foreign passenger was being very rude to my flight attendants on the ground, and he made it a point to offend each of the five attendants. According to the accounts I received, his behavior toward them was also offensive to the passengers around him. He finally said something to one of the younger ladies that couldn't go unanswered, at least, in her mind. He said, "In my country, all you could be is a house servant." Without hesitation, the young attendant responded by saying, "Sir, in my country, all you could be is a cab driver." There is absolutely nothing wrong with being a cab driver, but this employment was a little low on the scale for this particular individual. With that answer, the man closed his mouth and made no further rude remarks. The situation was handled without my intervention, and I appreciated that.

One of the many things I loved about my job was to give people tours of the cockpit while on the ground. Some would have their picture taken in my seat while others had all kinds of questions. When parents brought their children to the front for a tour, I relished many of their comments. One young lad asked, "What is the most important button or instrument in this cockpit?" I told him that, without a doubt, the most important instrument was the attitude indicator. This is the instrument that permits us to keep our wings level without reference to the outside. Without this instrument, the airplane could easily encounter dangerous and unusual attitudes. This is one of the primary instruments when "flying by instruments" in all sorts of weather.

A pilot has to learn, early in his training, to recognize and respond to a bad attitude. With his attitude indicator he can always maintain a safe frame of reference in relation to the horizon. This is true either in straight and level flight or when making turns. It would be great if man would use the built-in attitude indicator he is given, which is his conscience. As I have observed people through the years, it is apparent that some people do a better job of "flying by instruments" than others. Compare passengers like Ronald Reagan, Bob Hope, Ethel Waters, and many others, to the likes of the man above who has to see how many people he could insult. How good is your attitude when things don't go according to your plans? I once read a story about Zig Ziglar, the famous motivational speaker, who was traveling from Kansas City to Dallas. When told that his flight had been cancelled, he responded by saying, "Fantastic." The ticket agent looked at him in disbelief because she had been dealing with some unhappy passengers. He said, "Ma'am, there are only three reasons why someone would cancel that flight. First, there must be a problem with the airplane. Second, there must be a problem with the crew flying the flight. Third, weather could be a problem. Ma'am, if any one of these three problems exists, I don't want to fly either." I will add one more to his list: there might not have been an airplane available. Mr. Ziglar was making the best of an unfortunate situation. We can do the same.

A few months before I retired in July of 2001, I had the privilege of having former President Jimmy Carter on a flight from Seattle to Atlanta. We were flying an all night flight that was scheduled to arrive in Atlanta early the next morning. When I arrived at the airport, I was greeted by a ticket agent who advised me of my special passenger. When you carry someone of this importance, you are going to be scrutinized closely. The Secret Service and the FBI boarded the plane to make sure everything was safe. When President Carter boarded the first class section, he was accompanied by four Secret Service body guards. One sat next to him, and the others were dispersed throughout the first class section where they could watch him at all times. These body guards had to follow very strict rules. Only one could be asleep at any given time.

Before taking his seat, President Carter came up to the cockpit to say hello, and tell us how much he enjoyed flying on Delta. In our discussion, I asked him if he was still teaching Sunday school. He told me that he was teaching the Book of Ezekiel. I inquired why he chose such a difficult book to teach, and he responded by saying that his wife, Roselyn, posed a similar question. She said, "Jimmy, are you going to teach this where I can understand it?" I didn't inquire about his response, but I assumed the answer was yes.

Having Jimmy Carter on my flight that night was a little ironic. The previous night I was in Boston, and President Bill Clinton was giving a speech at the Park Plaza Hotel where I was staying with my crew. The irony was how these two people who held, or had held, the highest office in the land could have such differing moral perspectives. I told President Carter about President Clinton being in Boston the previous night. I also told him that while I did not always agree with all of his (President Carter's) policies and positions when he was president, I did appreciate the fact that he set a positive moral example.

During the flight, I mentioned to my first officer that I should have said something to President Carter about an interview he had when he was running for president. He was asked if he had ever cheated on his wife, and his response was that he had lusted in his heart. He received a lot of static and criticism from the press regarding this statement, and I wanted to tell him that I appreciated the fact that he didn't let it go past that. As I left the cockpit to use the restroom, the Secret Service agent who sat next to him was leaving the restroom. He stopped to talk and I mentioned to him that I wanted to ask President Carter about the "lusting in his heart" incident, but I missed the opportunity. He admitted that he had also "lusted in his heart," and I told him I was sure that his wife appreciated the fact that it did not go any further. I mention this because I have a good friend who told me I had to have some stories involving sex in my book if I wanted it to sell. This, along with a couple of other minor stories, is the extent of the sexual content in my book.

When I prepared for a flight, I conducted a preflight inspection to insure that the plane was okay. I also introduced myself to the head flight attendant and briefed her about what I expected during the flight. On one such flight, after my briefing, the attendant announced that we had Jesus on board. There was a passenger by that name. I went to the cockpit and informed my flying partner that we had Jesus on board. He said that he needed to go back there and meet him. I told him that I had already met Him (meaning the real Jesus) and he said, "Have you met him face-to-face?" I said, "No, but one day I will meet Him face-to-face." The copilot responded by saying that he hoped he would one day meet Him that way also. I had the opportunity to tell him that he could have the assurance of one day meeting Christ face-to-face. He needed to change from a feeling of just hoping to a confident assurance.

On December 15, 1982, the employees of Delta gave their company an airplane as a gesture of appreciation for exemplary personal treatment. This airplane was named "Spirit of Delta." As I flew this plane on a regular flight from Atlanta to Salt Lake City, one of my flight attendants said, "Are you in the spirit, Archie?" I told her that I was in the Spirit, and the Spirit was in me. I prefer the Spirit I'm referring to over the one to whom she referred. I have a picture of the "Spirit of Delta" hanging in my den. It has a piece of the red ribbon that was wrapped around the airplane when the employees gave it to their company.

One late night, when most of the passengers were asleep, the copilot and I were discussing our views of religion. In the middle of our discussion three flight attendants came to the cockpit at one time. They wanted to know our topic of conversation. My partner told them we were trying to solve the world's problems by religious means. One of them asked him what religion he was associated with, and he told her he was a non-practicing member of the Church of Jesus Christ of Latter Day Saints. Another flight attendant asked me if I had any religious affiliation. I told her that I did, and the church I belonged to was the Church of Jesus Christ of Present Day Saints. My copilot almost choked on his cup of coffee. This answer paved the way for a wonderful time of conversation and sharing about the Gospel of Jesus Christ.

I never knew what was going to happen in my job. The flight from Jacksonville to Atlanta is less than an hour, but a lot can transpire in that amount of time. I was flying the airplane and my first officer was talking on the radio to air traffic control and making most of the announcements to the back of the plane. Once we climbed to our cruising altitude, I received an unusual request. My head flight attendant came up to the cockpit with this bit of information along with her appeal. A man on board wanted to propose to his girlfriend sitting next to him. He wanted to have it done in a unique way, and he asked the flight attendant if she would speak to the captain to see if I would make the proposal over the public address system so everyone could hear it. I said that I would be happy to comply and told my copilot that I was going to relieve him of his responsibility of talking to the passengers for this one announcement.

The flight to Atlanta was less than one hour in duration so I didn't have a long time to think about what I was going to say, but I put a few words together and then made the pitch on his behalf. In my closing remarks, I added this request. "In a few short minutes we are going to be arriving in Atlanta, and I know that your companion would like an answer before we land." About three minutes after the proposal, the flight attendant made this announcement. "Ladies and gentlemen, we have a YES."

When I returned home to Salt Lake City, I said to Jody, "What would you say if I told you I proposed to another woman on this trip?" Jody looked at me and said, "Do I look worried?" I thank God that He gave me a wife who has complete trust in me. With the help of God, you too can make more of your moments memorable.

Boeing 737-200

In 1971, I transitioned from flight engineer to copilot. I moved from the Boeing 707/720 to the Boeing 737. In 1967, Boeing decided that the smaller, short-range 737 twinjet was the logical airplane to complement the 707 and the 727. There was increasing demand for transports in this category, but the 737 faced heavy competition from the Douglas DC-9 and the British Aircraft Corp. BAC-111. Boeing is still building this plane in advanced versions. I upgraded to the position of captain on this aircraft in 1978.

From Here to There

Allow me to "fly" you through a scenario of what a week was like for me in my life as pilot. This is a series of trips that I actually flew the month before I retired including specific details and actual times.

On this sequence of trips, I was scheduled to depart Salt Lake City at 8:35 a.m. on June 26, 2001. I had to leave my home in time to check in an hour before the scheduled departure time. It was my custom to be there just a little early as I was more relaxed with extra time to do the little things that needed to be done. One of those things was making sure my flight manuals were up to date. We would get approach plate revisions twice per month, which had to be updated before departure. Another important item was checking the weather for all the destinations I would be flying to that day, and examine the forecast for the next few days. After I completed some of these requirements and signed in for the flight, I was required to be at the airplane thirty minutes before departure. This gave me time to meet and brief the flight attendants, and complete my preflight inspection of the vessel I would be flying. I always allowed a little extra time for these requirements.

This preflight inspection consisted of checking all of the required equipment, while correcting any problems which the plane may have had when it arrived by checking the log book. After this was accomplished, the copilot and I settled into our seats to get ready for departure. We began by reading the "before start" checklist. Most of the items on this checklist were already done, and the checklist is a backup to make sure the things were done correctly. When flying for the first time with a pilot, I conduct a briefing to let him, or her, know what I expect for the next few days. As everyone does things a little differently, it is a good idea for the two pilots to be in agreement on how they will work together. I required that our jobs be done in the same manner in which we were trained to do them. My first consideration was for safety and this was the only reason that I would ever deviate from established procedure. When any sort of problem came up, it was always great to be able to say, "I did the job as I was supposed to do it."

On this day, were departing to Dallas/Ft. Worth for the first leg of our flight. The agents finished boarding the plane, and all the passengers took their seats. When this was accomplished, the flight attendant in charge reported that the cabin was ready for pushback. I contacted ground control for clearance to pushback and after receiving this clearance, reported to the tug operator that the brakes have been released. As the airplane started to roll, and when we were in the clear, we were given the "all clear" to start the engines. By the time the tug operator finished the push we had the engines started and were now ready to begin our taxi. Once the man waved us off, we reported to ground control that we were ready to taxi. He gave us instructions for proceeding to the active runway which included the taxiways we were to use. While we taxied to the runway, the copilot read a few more checklists and each of us responded to the items that pertained to us. In the cabin, the flight attendants prepared for departure by making sure that all the passengers had fastened their seat belts, seats were in the upright position, and they made safety announcements. As we approached the runway, we made an announcement to inform passengers that takeoff was imminent. The crew in the back of the plane informed us that they were ready.

At this time, we switched our radios from the ground controller to the tower operator. When we reached the number one position and the runway was clear, we were cleared onto the runway for takeoff. All of the checklists were completed and we were ready for the big event. We then lined up with the center of the runway and the throttles were advanced toward the takeoff position. The airplane began to roll, and we headed toward the sky. During the takeoff roll, we monitored the operation of the engines to make sure they were working properly. In the meantime, our speed increased and we reached the point where we started to rotate and become airborne. On the first flight segment of a trip, I usually performed the takeoff and landing and then alternated with the copilot through the remaining segments.

Immediately after liftoff we retracted the landing gear, but the flaps remained in the takeoff position until we reached an altitude of one thousand feet. Upon reaching one thousand feet above the airport elevation, we "cleaned" the aircraft by retracting the flaps. At this time I could use the autopilot, but when I flew I preferred to fly the machine by hand until I reached a certain point in my climb. I then insured that the plane was flying at the correct speed and altitude. From the time of our takeoff, our speed is increased to the desired speed of two hundred fifty knots until we reach ten thousand feet. We then climbed to our assigned altitude. The cruising altitude was determined by several factors, such as weight, weather (including wind and turbulence), and other airplanes in the vicinity. Upon reaching cruising altitude, we thanked the passengers for flying with us and gave them information about the route of our flight. We informed them of our expected arrival time, gave a weather report, and advised them of possible delays due to weather.

The crew and I would have time to sit back and relax for a time of conversation. We continually monitored the performance of all the systems on board, and we stayed abreast of weather patterns at our altitude in order to make the flight as smooth and comfortable as possible. If significant thunderstorms were in our path, we used radar, sought advice from the air traffic controller, and consulted other aircraft ahead of us on the same route. In other words, we used every resource at our disposal to give our passengers the best possible flight. Even though we used all of these different methods, there were times when we had to just ride out some rough air. Flying is like life. There are times when you encounter a little turbulence. It is not always a bed of roses, but you can overcome the rough spots if you keep your focus on the prize.

The scheduled flight time from Salt Lake City to Dallas was two hours and forty-four minutes which included everything from the time we started our pushback to the time we shut our engines down at DFW. The actual time in the air was just over two hours. As we cruised, the flight attendants kept us posted of things pertaining to the passengers, and we accommodated special needs or requests when possible. Sometimes medical concerns arose and we took whatever action was necessary to solve the problem. If the problem was serious, we called ahead and requested assistance. Once in a while we landed at the nearest suitable airport to obtain the needed medical treatment. In my career, I had only a handful of situations where I had to summon medical help on arrival, and only made one unscheduled stop for this reason.

The flight had gone smoothly, and we began our descent into Dallas. We had obtained the latest weather report and knew the runway on which to land. We had reviewed our applicable approach plates in order to program the computer with the required information. We were getting close to the most enjoyable part of a flight for a pilot: landing. During the approach and landing, we remained aware of possible changing weather situations and other planes in our general area. The preparations had been accomplished, and we completed the approach and landing. As we grew closer to the runway, we reduced our speed to get into a landing configuration. This meant that we needed to extend the flaps which are called "high lift devices." These devices permit us to maintain our lift at slower speeds. During the last part of the approach, we extended the landing gear and put the flaps in the landing position.

Extending the landing gear down prior to landing makes for a smoother touchdown. I know that I have made a few landings where people wondered if I had the landing gear down. I picked up an airplane in Atlanta one day, and the pilot mentioned that his landing was a little firm. Some pilots give more information in their announcement to passengers than others. Some give the weight of the airplane, the speed at landing, the weather, the elevation of the airport, and other pertinent items. After one such announcement about a hard landing from the captain, I heard a passenger say to another, "I wonder what the elevation of Atlanta is now after that landing." With that comment, I assumed that it was rather firm. I had a few firm landings in my career, but I don't think I ever altered the elevation of the airport.

After we landed the airplane and slowed down to taxi speed, we exited the runway and proceeded to the gate. We had received the gate assignment before landing and reported this to ground control once we left the runway. They gave us directions to the gate, and upon reaching it, we shut down the engines and the gate agent opened the door for passengers to deplane. The pilot that made the takeoff and landing for that leg of the flight bids the passengers goodbye. On this particular flight, we were in a Boeing 767 that carried over two hundred and fifty people.

The next few legs of our flight are a good illustration of a typical four day trip. After the passengers had exited the plane, we prepared for our next flight. We had forty-five minutes on the ground in Dallas in which to take our flight kits and suitcases to another gate to prepare a 767 for a flight to Jacksonville, Florida. We met a different set of flight attendants who were with us for the next two legs of our journey. The flight to Jacksonville was scheduled for two hours and twenty-five minutes. We performed our duties, and the flight proceeded without a hitch. In Jacksonville we had an hour and twelve minutes before proceeding to Atlanta and remained on the same plane.

We were enroute to Atlanta with a flight time of one hour. We had help with the wind and arrived twelve minutes ahead of schedule which made the passengers happy. In Atlanta, we changed airplanes again for the flight that took us to Dulles International Airport outside of Washington, D.C. We flew a Boeing 757 with over one hundred and eighty passengers. The flight time was supposed to be one hour and thirty-three minutes but we arrived fifteen minutes ahead of schedule. When we finished our duties at the airport, we had been on duty for almost thirteen hours.

We completed our duties and met the driver who transported us to the hotel. Since we were to fly from Ronald Reagan Airport the next day, we spent the night near that facility. That meant we had about a forty-five minute ride to our resting place for the night. Arrangements for our accommodations were made in advance by our company, which made checking in very easy. We had only to sign our name and pick up the key.

We had spent a comfortable night in Washington and were ready for a much shorter day. Our flight was not scheduled to depart until 5:07 p.m. and we had most of the day to do some sightseeing. I caught the Metro and went to the National Air and Space Museum near the Capitol. I also took advantage of seeing a couple more museums that make up the Smithsonian Institution. I have always enjoyed my visits to this place. After I had done my part of trying to impersonate a tourist, I returned to the hotel and prepared to do the job for which I am paid. We were scheduled to fly from Ronald Reagan Airport to Atlanta and then to Kansas City where we were to spend the night. The funny part of my employment is that I never really considered it a job. I felt as if I were on an extended vacation and truly enjoyed the opportunity that God had given me.

On the flight to Atlanta we carried the Prime Minister of Uganda, Apolohonpro Nsibambi. Accompanying him were with his daughter, Rhodamrs Nsibambi, his body guards, and Secret Service agents. After they took their seats, I invited them to the cockpit and let them sit in my seat while I spoke with him about his education, career, and a number of other things. I took a picture of him in my seat that he only allowed once I assured him that I would not show his feet. He said, "Archie, I will permit you to take my picture if you don't have my feet in it." He had left his shoes in the first class cabin.

The flight to Atlanta was scheduled for one hour and forty-one minutes and we beat this by eight minutes. We spent about an hour and thirty-seven minutes in Atlanta before proceeding to Kansas City for a one hour and fifty-four minute flight, arriving fourteen minutes early. On this day, we were only on duty for six hours and forty-four minutes. Our stay at the hotel was shorter than the previous night, but I still had plenty of time for a nice walk through the city and to enjoy a leisurely breakfast in a quaint little cafe`.

After completing our stay in Kansas City, we departed for Atlanta, Jacksonville, back to Atlanta, and then Orlando. We were on duty just shy of eleven hours on this day and our stay in Orlando would be only a short twelve hour stay. Early the next afternoon, we departed for Dallas/Ft. Worth and Salt Lake City, and we arrived home in Salt Lake City at a few minutes before 6:00 p.m. on June 29, 2001, completing a typical four day trip. The trip kept me away from home a total of eighty two hours and thirty-nine minutes and we accumulated twenty-three hours and fifty-three minutes of flight time. There were twelve takeoffs and landings, and we were on duty for almost forty hours.

Flights such as those mentioned above provided me with many wonderful moments. I met people from all walks of life. Flying out of Washington, D.C. afforded me the opportunity to speak with many different people with interesting experiences and perspectives. These people were placed in positions where they made decisions which impacted us all. We often had members of the United States Supreme Court, senators, congressman, governors, lobbyists, and people in numerous other occupations as well. Wouldn't it be great if all the people in these important positions based their decisions on their dependence on God? In the history of America, there was a time when this was true, and I would like to see America return to this dependency. Many of the decisions being made in Washington and elsewhere are being made by people who listen to the wrong message. When a person fails to consider God in his thinking, the decisions he or she makes are often for selfish reasons. If one's conscience is not in tune with the Creator of the universe, serious consequences can occur that negatively impact a lot of people. I asked David Souter of the United States Supreme Court how he could sleep at night when he had to make a decision that impacted so many people. He said that he did his utmost in order to give the best possible ruling on any particular matter. In my opinion, Justice Souter is a liberal member of the United States Supreme Court. He voted to uphold *Roe v. Wade*, a woman's right to choose abortion, in June of 1992. In still another ruling, he proclaimed that Boy Scouts should be required to accept gay scoutmasters in June of 2000. In June, 1998, he voted that felons may possess guns unless a state explicitly prohibits it. I pray for members of our judicial system who will uphold the laws of God without compromise.

Boeing 727

In the mid-1970s, I transitioned to copilot on this aircraft and flew in that position until 1978. The 727 was designed to service smaller airports with shorter runways than those used by the 707s. Of all the early Boeing jets, the 727 had the most distinctive appearance, with its rakish T-shaped tail and its trio of rear-mounted engines. It carried billions of passengers on everything from short hops to cross-country flights. The first 727 rolled out of Boeing on November 27, 1962.

The 727 was the first Boeing jetliner to undergo rigorous fatigue testing, the first to have completely powered flight controls, the first to use triple-slotted flaps, and the first to have an auxiliary power unit (APU). The APU was a small gas turbine engine that eliminated the need for ground power or starting equipment in the more primitive airports of developing countries.

Going Straight to Heaven

I read an article in *The Salt Lake Tribune* a number of years ago that dealt with the reason why male members of the Mormon Church lived longer than men of other denominations. The article was written using an address given by the leader of the Mormon Church in this town. Shortly after reading this article, I had the opportunity to fly with a man from that church. I asked him why he believed that men of his faith lived longer than men of other denominations. He said that he didn't have a clue and asked for my opinion. I told him that the reason could be that men of some other faiths are more anxious to get to Heaven. He registered a muffled chuckle at my response.

In August of 1995, I traveled from Atlanta, Georgia, to Tampa, Florida. My first officer flew the plane while I spoke on the radio. We received clearance along a certain jet airway between Atlanta and Tampa at an altitude of thirty-five thousand feet, or flight level 350. This is comparable to a highway roadmap with intermediate points on the route, which are called intersections. On this particular airway, there is a place called "Heaven." The actual designation is **HEVVN** because only five letters can be used to designate an intersection.

My coworker was an active member of the Mormon Church. Because of the interest that we both had in spiritual matters, we had participated in some wonderful dialogue regarding our thoughts about how to get to Heaven. While we shared similar thoughts, we also had major differences. We were blessed with the chance to voice our differences without becoming upset with the other's viewpoint. This is not always the case when discussing differing thoughts about religious matters.

While he flew the airplane, I asked him if he would like to go straight to Heaven. He knew what I meant because he was aware of the position of the airplane along the airway. He answered in the affirmative by saying that he would like to proceed straight to Heaven. I made the request to the air traffic controller that we would like to go straight to this intersection, and we were cleared upon leveling at flight level 350. When I made the request, we were still climbing to our cruise altitude. Please bear in mind that I would not want to get on the public address system of the airplane and ask my passengers if they wanted to go straight to Heaven. A few nervous passengers may misinterpret my motives.

After receiving the clearance, with a restriction of being level at the assigned altitude, I asked my copilot if he understood the significance of the clearance. He had a puzzled look on his face, and asked if I meant that you had to be on the level to get to Heaven. I said that was not exactly what I had in mind. For us to get to that ultimate place on an individual level, God also places requirements on us. We must first admit that we are a sinner and that we cannot save ourselves, and we must believe that Christ is Who the Bible says He is.

> For all have sinned, and come short of the glory of God;
>
> Romans 3:23 (KJV)

> That if thou shalt confess with thy mouth the Lord Jesus, and shalt believe in thine heart that God hath raised him from the dead, thou shalt be saved. For with the heart man believeth unto righteousness; and with the mouth confession is made unto salvation.
>
> Romans 10:9-10 (KJV)

My copilot said that he finally understood what I meant, but just believing wasn't enough. According to him, one must do good works in order to be admitted to that Kingdom. I reminded him of the following passage.

> For by grace are ye saved through faith; and that not of yourselves: it is the gift of God: Not of works, lest any man should boast.
>
> Ephesians 2:8-9 (KJV)

He quoted a passage from James back to me.

> But wilt thou know, O vain man, that **faith without works** is dead?
>
> James 2:20 (KJV)

This passage of Scripture has become a major stumbling block for numerous people and religions. It is my belief that an active faith will produce good fruit, but our works do not provide our salvation. If this were true, we could not say that our gift from God was free. Christ fully and completely paid the price for our sin. The fruit produced in our life is because of our profound love for the One who gave so much for us. When faith and works come together harmoniously, we show evidence of our faith. This is how we can have an impact on the world around us. The following is an illustration that I have seen and heard in several different versions.

An old boatman painted the word "faith" on one oar of his boat and "works" on the other. He was asked why he did this. In answer, he slipped the oar with "faith" into the water and rowed. The boat, of course, made a very tight circle. Returning to the dock, the boatman then said, "Now, let's try 'works' without 'faith' and see what happens." The oar marked "works" was put in place and the boatman began rowing with just the "works" oar. Again the boat went into a tight circle but in the opposite direction. When the boatman again returned to the wharf, he interpreted his experiment in these strong and convincing words, "You see, to make a passage across the lake, one needs both oars working simultaneously in order to keep the boat in a straight and narrow way. If a person does not have the use of both oars, he makes no progress trying to cross the lake or in his walk as a Christian." Our discussion continued in the cockpit without either of us convincing the other to change our position. I am not responsible for the decisions someone makes, but I am responsible for sharing my faith with others when God provides the time and place.

A few days after returning home from this trip, I received a letter from this pilot. It was a twenty-eight page explanation of why he believed what he did, plus a listing of every religious book in his home library. I responded to his lengthy letter with a one page reply. I included the statement that twenty-eight pages are not needed to explain how to get to Heaven. It can be summed up in three words. Repent and believe. When he received my letter, he answered with a twelve page letter. He was confident in his belief, but in my opinion, he was dead wrong.

Some years later, a flight attendant who was the wife of this copilot was part of my crew. I asked her if her husband had ever told her about our trip to Heaven. Her husband had never told her of this instance, but she said she couldn't wait to hear the story. After completing her assignment in the cabin she returned to the cockpit, and I told her the story. She was thrilled that her husband had made it to "Heaven." Don't take that in the wrong way.

There was another time when I walked by Heaven. On July 14, 1998, I was out walking in New York City on a layover. It was hotter than hell, and I saw GOD coming toward me. Let me explain.

I was on Lexington Avenue, several blocks from the hotel where I was staying. I was near a sign that said, "Heaven is only sixteen floors above ground," and it pointed to a restaurant called the Fountain Room. It was a very hot, muggy day, but I was out enjoying my walk. Suddenly a truck drove toward me with the letters GOD painted on the front and sides. The letters were very large, and they certainly got my attention. Here it was hotter than hell. I was passing by Heaven and GOD comes along. That is enough to get anybody's attention. As the truck approached, I became curious about what GOD represented. When it passed by, I saw that GOD stood for "guaranteed overnight delivery."

How does this apply to my life? When it is very hot and muggy, it is not the most comfortable situation. There are times in my life when I have the desire to be out of certain situations that I am going through. What do I do in those situations? I'm afraid that too often I start the exhaustive process of climbing those sixteen flights of stairs to get to Heaven, and I realize that those stairs can be mighty steep. Because I have been saved by grace through faith by the shed blood of Jesus Christ, I suddenly realize that I don't have to climb all those stairs to reach Heaven.

> Neither shall they say, Lo here! or, lo there! for, behold, the kingdom of God is within you.
>
> Luke 17:21 (KJV)

> And saying, Repent ye: for the kingdom of Heaven is at hand.
>
> Mathew 3:2 (KJV)

When I come to this realization, it is as if someone turned on the air conditioner and the heat of my problems start to melt away. I don't have to depend on a truck with GOD on the side that promises overnight delivery. Knowing that the Holy Spirit was within me, and I didn't have to wait until tomorrow to be delivered by Him, it just seemed a lot cooler. This permits me to see beyond the problem for the moment. Don't keep the Holy Spirit locked in an obscure room out at a distance from your everyday life. Stop and remember that He is within you ready to meet and supply all your needs.

The Road of Life

Down the road of life there's a bump or two
Sometimes it seems like I'll never get thru.
When they get the roughest there's a detour around,
So I don't have to let those bumps get me down.

God gives me grace to handle those bumps in the road,
But sometimes I struggle as I carry the load.
Then I remember how to take my load and make it trim,
All I have to do is take it and give it to Him.

A. B. Lawson

We sometimes have to look at things from a spiritual perspective. In a sermon, a preacher told a story about a couple and their four year old daughter who went on a cruise. As the ship pulled away, the passengers stood by the railing and waved their goodbyes to those on the dock. The little girl stood behind her father and could not see anything. She tugged on the leg of his pants until she got his attention and said, "Daddy, I want to see also." Without hesitation, the father hoisted his daughter to his shoulders so that she towered above the others. As she gazed toward the water and the shore, she made the remark, "Mommy and daddy, I can see farther than my eyes can look." If we try to look at things from a Biblical perspective, we can sometimes see farther than our eyes can look. God promises wisdom if we only ask Him for it.

There is one more story regarding the subject of Heaven. On November 7, 1997, my copilot and I flew from Las Vegas to Salt Lake City. We were about twenty five minutes from landing when we were informed that we were going to be parking at gate C8.

Upon being informed that this would be our gate, my copilot said, "I guess things are going rather well for me in this life because I am able to park at gate C8." He said this because we were getting home seven or eight hours ahead of our scheduled arrival time, because part of our trip had been cancelled, and we were assigned to a rather convenient gate. The company changed our original schedule for operational reasons, which happened sometimes because of factors ranging from equipment changes, weather, and availability of crew members. I told him that I was looking forward to parking at gate C1 in the next life, and I proceeded to tell him why. In my story, C1 represents Heaven.

Gate C1 is the best gate at which to park when ending a trip because it is closest to the pilot's lounge, and you therefore don't have to carry your flight bag too far to store it before going home. It is the shortest walk. When it comes to gates, it doesn't get any better than C1 at the Salt Lake City International Airport.

My copilot said, "I feel like I am always destined to get gate D13." This is the gate that is the farthest distance from the pilot's lounge. I then asked him what gate he expected to get in the next life. He said that he would feel fortunate to get gate D13. I told him he needed to reevaluate his religious beliefs. My copilot was a member of the Mormon Church, but he got to hear about the benefits of parking at a better gate. He will never be able to pass by gate C1 without recalling this conversation.

When I takeoff on my final flight and park my aircraft for the last time, I am going to be parking at gate C1. I can say this with assurance because the Word of God tells me so. Before I continue, I want to make it clear that I love the members of the LDS Church. However, we have theological differences in a number of areas. In my opinion, complete assurance of Heaven is missing from Mormonism. I have asked numerous people of this faith if they knew if they were going to Heaven when they died. The answer has always been essentially the same. "I would like to think that I have been good enough to get there." If I understand this correctly, this tells me that the individual is unsure that his "works" are sufficient to permit entry into Heaven. If this is true, I pray that they will one day have the same assurance that the Word of God promises. We're not going to get a ticket to Heaven because we have been good enough. We can only expect to get there if we have repented of our sins and accepted Jesus Christ as our Savior. I thank God for letting me know that He has the very best parking spot waiting for me when I fly on my last trip in life. Do you have that same assurance? If you feel like your destiny is gate D13, you can make your request to park at C1. All you have to do is call upon God and make your request. He is on call twenty-four hours a day, seven days a week, and three hundred and sixty-five days a year. How can we know that He will give us good advice? Let me tell you how I answered on one occasion.

> These things have I written unto you that believe on the name of the Son of God; that ye may **know** that ye have **eternal life,** and that ye may believe on the name of the Son of God.
>
> 1 John 5:13 (KJV)

In April, 1999, I flew nonstop from Boston to Salt Lake City. My copilot and I discussed religious matters when he asked me this question, "How do you know which bible is correct?" He was referring to the bible for the Muslims, Buddhists, Mormons, and the Christian Bible. I said, "Let's talk about things that are reliable and trustworthy. Would you say that this airplane we are flying is safe and trustworthy?" He answered, "Yes." I said, "This airplane has been thoroughly examined and found worthy to be used safely on this flight. As a matter of fact, the company we work for has had thorough examinations and their airplanes have been found safe for flight. They have been scrutinized many times over the years. This is not true of all airlines. Just recently an airline was grounded because its planes were found to be unworthy to carry passengers safely."

I continued, "I consider the Bible, God's Holy Word, to be like this airplane we are on. (In this discussion, I was referring to the King James Version.) It has been examined thoroughly and found to be without fault. Some of the other 'bibles' of other religions have been examined and have not stood up to the test of the Christian Bible. You wouldn't want to get on an airplane that has been examined and came up lacking. If I take a trip I want to go with the most reliable means available. As I go through life, I am on a trip that will eventually lead me to Heaven, and I want to rely on the most trusted equipment I can find. The Christian Bible gives me complete guidance and instructions for my journey. I don't want to depend on something that just might leave me short of my destination."

When it comes to deciding which Bible to read, one has many choices. Cliff Notes Bibles are available for people who don't have time to read the full version. How would you like to have a reservation on one of my flights if I told you that I had made my preparations for the flight using instructions from Cliff Notes? This may offer an overview of the entire Bible, but it only gives you a glimpse. I am looking for the Main Event, and I am not satisfied with a preview.

Boeing 727 in Delta paint scheme

I also flew this aircraft for Delta in the position of captain until late in 1990.

I Believe in Miracles

Why do I believe in miracles? Well, I'm glad you asked this question because I want to tell you why. In 1975, I was diagnosed with bronchiectasis, a condition where you are susceptible to bronchitis and pneumonia. The doctors told me that I would only be able to work from one to ten years and would not be working as a pilot when, and if, I reached the mandatory retirement age of sixty. The doctors believed my condition would deteriorate to the point where I would need supplemental oxygen for breathing by the time I was fifty. This condition did not affect my ability to fly. The only restriction that I had was not to fly when I was sick, which is exactly the same restriction for any pilot. This was a big wakeup call for me, because this diagnosis precipitated my taking a closer look at my mortality. I also became more aware of the importance of planning for my future in more ways than one.

In March of 1983, I contracted pneumonia. This was not an ordinary case. I hemorrhaged in my lungs. In addition to the problem of bleeding, my blood pressure was very high. Because of this trouble, I was forced to take a medical retirement. I was faced with the possibility that I would never work again at anything, and I was told to take "flying" out of my vocabulary. At that time, I was under the care of four doctors. One of the four doctors wrote my letter for medical retirement, stating in the letter, "What Archie needs is a priest, not a doctor." They didn't know what the prospects for my future were, and I was told that I needed to adjust to the fact that my condition was going to continue on a downhill pattern. For a period of time, I was considered totally and permanently disabled. In addition to all of this, I was living on some rather powerful medication to ward off bronchitis and pneumonia.

I had wondered many times throughout my career how I would handle having to stop doing the thing that I so much enjoyed. There was an earlier time in my life when I would have been devastated because flying was almost like a god to me. This perspective began to change after I became a Christian in 1969, but it was certainly not an immediate transformation. By the time I had to take a medical retirement in 1983, flying had moved down the list of important things in my life. My relationship with God was first, my wife and family were second, and flying occupied the third spot on the list. When I was told I couldn't fly anymore, I still had the two most important things in my life. Flying was hard to set aside, but I was not devastated emotionally by it.

For the first couple of years after I retired in 1983, I was in rather poor condition. Jody instituted a number of dietary changes to try to assist me in changing my condition for the better. I also went through a long-term supervised fast. This had a remarkable effect on my physical state, but I still was required to use a number of different antibiotics.

Near the end of the summer in 1988, I mentioned to Jody that I felt good enough to try to regain my medical certificate in order to get my job back. I went to the doctor and told him my ideas about returning to flying. Once again, he said that I should just go fishing and remove the thought of flying from my mind. I requested that he give me the toughest physical imaginable to see if a possibility existed for me to return to flying. He agreed and gave me a physical that lasted over three hours. After finishing all the elements of this examination, the doctor said that he could find no reason why I couldn't return to flying if I so desired, but his personal advice was that I not fly.

I mulled over his advice for the next few days, and decided to go back to the FAA medical examiner to see if I could be reinstated. I passed the exam with the FAA, and could now return to the job I loved so much. After I passed these medical tests, my mother-in-law wanted to know who deserved the credit. When my medical history was examined and the doctors interviewed, one would have to conclude that it was a miracle to be able to return to flying. I was not ready to claim a miracle at this time, because I knew there was a possibility that I could encounter this problem again. When you consider that the doctors told me that I would carry an oxygen bottle if I lived to be fifty, most people would call this a miracle in a second. Let's wait a while before we label this a full blown miracle. I said to my mother-in-law, "If I go back to work and fly for six months without a problem, then I will put the full miracle label on it." In my mind, the timing was not quite right. There would be a time very soon, however, that I would have no regrets for making this statement.

> And it happened, as He was alone praying, that His disciples joined Him, and He asked them, saying, "Who do the crowds say that I am?" So they answered and said, "John the Baptist, but some say Elijah; and others say that one of the old prophets has risen again." He said to them, "But who do you say that I am?" Peter answered and said, "The Christ of God."
>
> And He strictly warned and commanded them to tell this to no one, saying, "The Son of Man must suffer many things, and be rejected by the elders and chief priests and scribes, and be killed, and be raised the third day."
>
> Luke 9:18-22

Perhaps you wonder how this passage relates to my situation. I believe, beyond a shadow of a doubt, that God allowed me to return to flying to bring glory to Him by sharing with others what He did for me. When the Disciples of Christ responded to His question of, "Who do you say that I am," Peter answered and said, "The Christ of God." Christ then warned them to tell this to no one. At first, this statement seems somewhat puzzling until you understand that His testimony would not be complete until after the resurrection. The timing was not quite right. How does this apply to me?

When I medically retired in 1983, there were some who believed that I faked my physical illness because Western Air Lines was having financial problems at the time. Some were able to get a doctor to certify that they were physically unfit for flying, which allowed them to take a medical retirement and receive disability benefits, even though they still met all the physical qualifications. The people who felt I was in this group were not fully aware of my condition, but this did not matter because some are going to believe what they want. In my case, there was something else that needed to happen before my credibility in this matter would be restored. Then, and only then, could I share the testimony of my miracle and have people believe. At that time, I did not have a clue about what was to come, but I will say that all doubts regarding my miracle would soon come to an end.

In December of 1988, I was back in the classroom going through the process of renewing my qualifications. Even though I had over five thousand hours in the Boeing 727, I still had to go through the complete school because of the amount of time I was away from flying. The re-qualification process lasted about six weeks. In the middle of January of 1989, I successfully finished all of my training. As I prepared to go back home to California, I made a trip to the restroom. This was not going to be a very pleasant detour. While in the restroom I cleared my throat, and began hemorrhaging from my lungs once more.

work for the second time, and had worked for six months, I had no doubt about the miraculous blessing God had given me, and told others about it

Since I had not flown for several years, a check pilot accompanied me as copilot for my first twenty-five hours. This is standard procedure and required by the FAA. The flight instructor was a friend with whom I had flown in past years. During the first few days we were together, he told me that some things in his life were not what he wanted them to be, and he felt that something was missing. His relationship with his wife and family was good, but a piece of his life was out of place. He knew that I was a Christian, so I walked him through the message of the Gospel. At the end of our little talk, I gave him a copy of the New Testament and gave him a chance to accept Christ. He said, "Archie, I know that I need to do that, but the timing is not quite right. I need to think about this before I take that step." During the next two years I saw him on a fairly regular basis. He was still waiting to make that decision. Near the end of those two years, he told me that he had developed a cough that repeated antibiotic treatments would not heal. He was concerned about the big "C" word. He went in for some further tests, and it was indeed confirmed that he had lung cancer. The doctors felt they had discovered it early, and they believed there was chance to successfully treat it.

He checked into Palo Alto Medical Clinic at Stanford University for a biopsy. When they operated on him for this, they discovered that the cancer was too far advanced to treat with surgery. They closed him back up and gave him just a few months to live. I made a trip to visit him at his home in the eastern California mountains shortly after he got out of the hospital. After we said our little hellos, our conversation was headed toward discussing mundane, unimportant things. After satisfying his desire to talk about politics, weather, and other things, I wanted to move our talks in a little different direction. I asked him if he recalled our discussion about Christ almost two years earlier. He had the New Testament close by that I had given him. I said, "What have you decided to do about Jesus Christ? You told me then that this was something you needed to do." He told me that he had just recently given his life to Christ, and I welcomed him into the Kingdom of God. The rest of our time together was joyful as we talked about things of substance. My friend passed away a short time later, but I was consoled by the fact that he had made the most important decision of his life: he turned his life over to Jesus Christ. I praise God for that.

After I had worked for over six months, I had no problem claiming my miracle. I continued to work until I reached the commercial airlines mandatory retirement age of sixty, in 2001. In the 1980s, I was forced out of work twice because of medical reasons, but I was forced out in 2001 for normal reasons.

About a year after I retired, Jody and I made a trip to San Jose, California. In addition to seeing friends, I wanted to see the doctor who had grounded me in 1983 (he wrote the letter stating that I needed a priest instead of a doctor). He was amazed to learn that I had returned to flying a second time, and was able to do so until I turned sixty. We had a wonderful reunion, and I had my picture taken with him in his office.

Knowing is Not Good Enough

I was in Los Angeles waiting for a flight home to San Francisco in the early 1970s. My trip concluded there, which was unusual because we normally ended our scheduled flights at our home base. The first flight I could catch was a DC-10, and it was not on Delta Air Lines. The story I am about to tell you may seem like there was a safety issue involved. At no time were the passengers on board this airplane in any danger.

I went to the airplane and introduced myself to the crew flying to San Francisco. Since I was in my pilot's uniform, they asked if I would like to ride in the cockpit with them. My intention was to rest in the cabin during the flight, as I had just completed a long work day. But I decided to accept the offer because I had not been in the cockpit of this type of plane in flight. We departed Los Angeles, and reached our cruising altitude with a little time to exchange pleasantries. After our scheduled landing in San Francisco, this crew was headed for Minneapolis, Minnesota, and would arrive quite late. It was already past ten in the evening.

We approached the point at which we must begin our descent for the arrival at the airport. Above ten thousand feet there is usually no restriction on the speed you maintain, unless air traffic control sees that you are getting too close to another plane. Below ten thousand feet the restriction is to slow to a speed of two hundred and fifty knots. In aviation we use knots instead of miles per hour. Knot refers to nautical miles per hour. The pilot did not slow down while descending below ten thousand feet and I began to wonder what he was doing. I have heard of people who have become impatient over the years and wanted to hurry things up a bit. A good reason for flying at slower speeds in lower altitudes is that, when in a congested area, there isn't much time to avoid other airplanes. The controllers also have to gauge the planes so that they arrive over certain points precisely in time to provide enough space the traffic ahead. This allows the airplane ahead to land and depart the runway before another touches down.

The plane was at an altitude of nine thousand feet and flying at about three hundred fifty knots, which was one hundred knots above the limit. Bear in mind that no one was in danger, but we were closing in on the runway environment and could not land at that high speed. If the pilot continued at this speed much longer, he simply would have had to overfly the runway. There were three pilots in the cockpit: the captain, the copilot, and the flight engineer, all at their respective control positions. The copilot was flying the airplane on this approach, and the captain and flight engineer never brought the speed of the plane to the attention of the man at the controls. I had difficulty keeping my mouth shut as I had made the decision to give them a chance to correct the problem, but finally approached my limit in self-restraint. From the standpoint of an observer, it appeared as if the captain and flight engineer were paying attention, but I soon realized that their attention was focused more outside the aircraft than inside. Both must be included in your scan pattern. The evening was very clear, and I suppose they were captivated by the beautiful night scenery. The lights of San Francisco and surrounding cities present an incredible view.

We continued our descent through eight thousand feet, to seven, six and then five thousand feet. I could no longer remain silent and asked the copilot why he was traveling at that speed at this low altitude. Without hesitation, the first officer retarded the throttles and extended the flight spoilers, or speed brakes. The airplane decelerated rather quickly, and we landed without having to fly past the airport. Why did the men flying this plane overlook basic principles in a descent? This cockpit held three highly trained pilots who without question knew the rules. This speed limit rule has been in force for many years. The point that I want to illustrate is this: you can know all of the rules and regulations, but if they are not put to use they will do you no good. The Bible has guidelines for every aspect of life, but if we are only a "hearer" and not a "doer" of the words contained therein it does us no good. We not only must hear, but we must be willing to do as well.

I once flew in the capacity of first officer for a man of another faith. We flew from San Francisco to Vancouver and the subject of religion came up for a few moments without anything of substance being discussed. This captain knew my feelings concerning his religious persuasion through friends who had flown with me. We had some time in Vancouver before going back to California, and he approached me and wanted to know if he could answer questions I might have regarding his beliefs. I assured him that I had no questions because I had a fairly good understanding of what those beliefs were. He insisted on a chance to enlighten me and I patiently declined to participate in this discussion. Once he realized that I didn't want additional information, he asked me to explain the main differences in our views. I responded with the statement, "Well the Christian view verses your view is this." He would not let me finish my statement and angrily said, "I'm getting good and tired of you people telling us we aren't Christians." I said I would be happy to give him my view if he could listen without getting angry. He apologized for his outburst, and we continued our discussion all of the way back to San Francisco. The flight engineer listened to our dialogue, and wanted to know if either of us had won the discussion. The captain said that I certainly didn't change his mind, and he acknowledged that he didn't change mine either. We ended the discussion with a smile toward each other.

The captain began the approach to the airport, and I backed him up and spoke on the radio. As he was about to turn toward the runway, he made a slight misjudgment in the turn that required input from me, but no safety issue was involved. After he made the correction and was lined up with the runway, I said, "Gerry, I was able to help you correct that little problem and everything worked out fine in this situation, but I won't always be there. If you come to the Day of Judgment with your beliefs and realize you made the wrong choice, it will be too late to say that I wish I had listened to what Archie had to say. We must make our choices before that time, because the Bible says that it is appointed unto man once to die and then the judgment." The flight engineer spoke up again and said, "I think I know who won the discussion." I don't think of this as a matter of winning. I think of it as trying to effectively share the Gospel with whoever will listen. I simply try to use the words God gives me for the situation and for the moment. The captain and I left on good terms. The way I look at this is that he has until that judgment day to accept the true Gospel. God is patient, but eventually that patience will end. Now is the time to respond. You may not have tomorrow.

During the early days of his ministry, Dwight L. Moody launched a series of meetings in Chicago that promised to draw the biggest crowds to whom he had previously preached. He spoke about the life of Christ, and on the first Sunday night, October 8, 1871, he chose a sermon that dealt with the trial before Pilate. As he came to the end of his message, he turned to Matthew 27:22, *"What shall I do then with Jesus, who is called Christ?"* He concluded, "I wish you would take this text home with you and turn it over in your minds during the week, and next

Sabbath we will come to Calvary and the cross, and we will decide what to do with Jesus of Nazareth."

Moody probably considered this a clever device to attract the attention of his audience. But speaking of it in later years, Moody called that conclusion to his sermon the greatest mistake of his life. Mr. Sankey was singing the final hymn:

> *Today the Savior calls;*
> For refuge fly;
> The storm of justice falls,
> And death is nigh-

The fire engines began to sound on the street as they made their way to their first contact with the great Chicago fire. Moody's hall was completely burned. It is estimated that over a thousand people lost their lives in the city wide fire. Moody never saw that congregation again, and some of those to whom he spoke on that night doubtlessly died. After this incident, Reverend Moody challenged people to decide immediately about Christ. Now is the time to respond to God's call. You may miss the opportunity if you wait until tomorrow. Even with all the warnings we are given, some believe they can make their decision tomorrow.

Early in my Christian life, Jody and I attended a retreat in Northern California. I was one of the discussion leaders. One subject in the series of meetings concerned the importance of responding early to the call of God on our lives. After one of the meetings, a young man approached me and asked, "Why can't I live the way I want to now, and just make my decision when the time is right for me." I told him that he was free to do exactly as he liked because God has given us the freedom of choice. My advice did not end there. I told him the following story.

Prior to this retreat, my crew and I were in Anchorage waiting to take off for Juneau, when we were advised that an Alaska Air Lines Boeing 727 had crashed in Juneau. The airplane had not been found. We were scheduled to fly on the first flight from Anchorage after that crash. A vice president from Alaska Air and their chief pilot joined us in the cockpit for the flight. The B-727 was found a few miles from Juneau just before we landed, and one hundred twenty-eight people died in that crash. I said to this young man, "Suppose that all of the people on that flight felt just like you. They were living just the way they wanted, but they had the intention of making a decision later." Their ability to make the choice at a later time was taken away in less than a second. I asked him, "Are you willing to take that chance?" This young man left the retreat without making a decision for Christ, but he left with more to think about than when he came. The choice was his as far as the decision about his future. Now is the time.

How Important are You?

> For I say, through the grace given to me, to everyone who is among you, not to think of himself more highly than he ought to think, but to think soberly, as God has dealt to each one a measure of faith.
>
> Romans 12:3

Never has there been a shortage of people who are looking for an opportunity to let you know how important they are. In addition to years of flying in the front of an airplane, I have done a considerable amount of traveling from the back as well. I have always enjoyed meeting and talking to new people. I have been blessed to be seated with many people who have provided me with countless hours of stimulating conversation. God has placed me in some great positions to allow me to receive a blessing, and I have had chances to bless others as well. At least, that is what some have said.

A great deal of importance is placed on airlines getting their passengers to their destination on time. On time performance reports are often published in the business section of newspapers and other publications. People are looking for reliability. What happens when things don't go according to plan? One scenario may be that you have an important meeting in another city, and it is imperative that you are there on time. You have been told that this meeting is the most important meeting your company has ever invited you to attend. Let me tell you about a couple of instances where plans had to be changed because of circumstances.

I have never understood how someone could board the wrong airplane with all the signs and announcements that are posted. Even though I don't understand it, this is something that has taken place from time to time. I should say that it is something that "used to happen." Today, your ticket is scanned before you board the airplane, and it will be rejected if you are trying to get on the wrong flight. People are usually aware of their mistake before the gate is closed and the plane departs from the airport. One such instance happened many years ago on a flight from San Francisco to Vancouver. A lawyer on board was to conduct a trial in Los Angeles. As he prepared intensely for the court proceedings, he was caught up in the moment and failed to hear the announcements made regarding the destination of this flight.

We were in the air, enroute to Vancouver, when he realized his mistake. He said to the flight attendant, "You mean this flight is not going to Los Angeles?" The flight attendant answered, "Sir, that is exactly what I have been saying." The lawyer emphatically said, "You must turn this airplane around and let me off so I can get on the flight to Los Angeles. I have to be in court later this morning." The crew member told him that he would not make it to court because we were going to Vancouver. We heard how we had ruined his opportunity to get to where he needed to be. There was no doubt in his mind how important he was. He made this point perfectly clear. According to him, we would have to pay for allowing him to board the wrong airplane. Dealing with this passenger and his situation left some of the other passengers waiting for service. Once we landed in Vancouver, this gentleman still had one more obstacle. Before he could board a flight and go back toward his destination, he had to clear Canadian Customs, as well as U.S. Customs. This left him in a much greater state of discontentment, and he had some time to consider how important he really was.

I have great respect for this man's right to get to his precise destination, and went out of my way to see that he boarded the right flight. Before the flight departed San Francisco, the gate agent came aboard and said. "If anyone has boarded this flight to Vancouver in error, now is the time to deplane." I then made my announcement of welcoming the passengers on our trip to this fair Canadian city. The flight attendants also made these announcements. If I had known his plight before we were airborne, I could have allowed him to get off.

Reasons why people don't arrive at their destinations according to schedule are numerous, and could include mechanical problems, weather, medical reasons, etc. Such a delay occurred on a flight from Seattle to Salt Lake City. I did not have time to take a survey to see which passenger had the most important appointment awaiting them on our arrival. The flight progressed normally, and in the vicinity of Boise, Idaho, the flight attendant called from the cabin and announced that one of our female passengers was having a seizure. I told her to make an announcement to see if we had a doctor on board, and three doctors responded. One of them was a critical care surgeon from the University of New Mexico who quickly came to the aid of this lady. I told the doctor that we could arrive in Boise in fifteen minutes or in Salt Lake City in forty minutes. He said that we must get the woman to Boise. From an altitude of thirty-five thousand feet, I declared a medical emergency, and we touched down in twelve and one-half minutes from the time we began our descent. I can tell you that my copilot and I were extremely busy during that short time.

Air traffic control was told to have medical personal standing by when we arrived at the gate. Sixteen emergency medical technicians boarded the airplane to assist the passenger. After she was taken to the hospital, we departed for the flight to Salt Lake City. Not one person complained about not arriving when they expected. I did have a couple of passengers tell me that they hoped I was their captain if they ever had an emergency. At this moment in time, this lady was more important than the flight schedule.

I stood by the cockpit saying goodbye to passengers as they deplaned, when one of the passengers said, "Archie, I didn't know you were on this flight." He was my Sunday school teacher from my church, and he often flew this route. After I finished saying my goodbyes, we had a few minutes to talk about the flight.

Several months later, I was flying this same route again. I was at the gate getting my paperwork from the gate agent when I bumped into this man from my church again. He was once again heading back home after conducting business in Seattle. We talked about the last time he traveled with me on this route. I took him to the airplane and gave him a tour of the cockpit. I introduced him to my copilot and flight attendants, and I told my crew to take care of him.

We departed on schedule with a completely full airplane of over two hundred sixty people on board. We also had a pilot deadheading to work with us in the cockpit. After we leveled off and headed to Salt Lake City, I told my first officer and the other pilot about the last time I had my Sunday school teacher on board. Shortly after finishing the story, we were once again in the vicinity of Boise. At nearly the same point as the previous trip, the other pilot said that the flight attendant had just made an announcement for a doctor. He was listening to the radios, and he had the button set to hear what the attendants were saying in the back of the plane. Usually I only turned that switch on when one of the crew members in the back called me. My primary responsibility was to focus my attention on air traffic control and monitor the progress of the flight.

The copilot and I were incredulous and said to the other pilot, "You're kidding! That can't be happening." The bell rang for the flight attendant to enter the cockpit and she told me that we had a passenger who was possibly having a stroke. I opened a line of communication with the doctor who was assisting this male passenger to determine what he needed. By this time, he was walking the gentleman up and down the aisle. This sick man did not want to go to Boise. He wanted to get home. After a few minutes of examination, the doctor told me that he believed the patient was able to continue to Salt Lake. I made the decision to continue the trip without further interruption.

I approached my Sunday school teacher after we landed, and told him that I wasn't sure it was a good thing for him to be on my flight. He went home and told his wife about this, and she said, "I don't think I want to go on any of Archie's flights if this is going to be something that he makes a habit of."

In the 1990s, I was flying from Dallas to Fort Lauderdale, Florida. Upon leveling at our cruising altitude, a passenger decided to retrieve a piece of luggage from an overhead storage bin. When he opened the door, his suitcase fell on the head of the person sitting next to him. This accident resulted in a slight neck injury. The flight attendant appeared at the scene very quickly to investigate. Her primary responsibility was to determine if the injury was serious enough for me to take appropriate emergency action. After interviewing all those concerned, we proceeded on to our destination. In an instance such as this, the attendant is required to fill out a form detailing all the aspects of the incident. This includes statements from those involved and the witnesses. Since the passenger was complaining of a sore neck, he requested to see a medical technician upon arrival. When he made this request, the passenger who dropped the case assumed this was a precursor to a lawsuit. At this time, he became somewhat belligerent by making a very unwise statement. He said to the person sitting next to him, "If you think you've got a sore neck now, just wait until we get to Fort Lauderdale and I will give you a real one. Are you Jewish?" With this threat, the slightly injured passenger was a little frightened, and this was reported to me. Without hesitation, I passed the word to the alarmed traveler that there was no need to be afraid. I immediately radioed Delta in Atlanta and advised them of the situation, and I requested that law enforcement officers meet the plane when I landed. All I had to do was give them the seat number of the offending passenger, and they took care of the rest. When we arrived, uniformed officers met the airplane with a complete history of the person who made the threat. He did not receive the welcome he expected. This man made a simple mistake, and he could have averted any problems by showing concern for the one he injured. All he was concerned about was his own welfare. The other person was not important to him.

On yet another occasion, someone tried to let everyone know how important he was. We were making preparations to depart from Philadelphia to Atlanta, and a passenger demanded more than his fair share of attention at the gate. He was so obnoxious that he interfered with the duties of the flight attendants as they prepared the cabin for departure. I was advised of his behavior, and made a request for a special agent to come on board to talk with the offender. He was told that he would be removed from the airplane if he continued his previous actions. With the flight attendant present, the man assured everyone that he would act appropriately. We all agreed that it would be fine to continue to Atlanta with him on board.

The door of the plane was closed, and we backed away from the gate. Once disconnected from the tug, we began our taxi toward the departure runway. As soon as we started to move under our own power, the passenger started causing trouble again. I wanted to continue on to our destination without delay for the other passengers, but I had to be assured that it would be safe to do so by taking into account the actions of this "important" belligerent person. I taxied the Boeing 767 over to the edge of the ramp where I would not block any airplane behind me, and I told the ground controller about my situation with the unruly passenger. He said it would be fine to park there as long as I needed.

My copilot was a practicing attorney in addition to his job as a pilot. I asked him to go back and see if the man would give us assurance that he would cooperate in order to continue our flight. I told him to let the person know that if we departed and he caused further trouble, I would land the airplane somewhere between Philadelphia and Atlanta and have him removed by the authorities for interference with a flight crew. My copilot left his seat and went back to have his conversation with this individual. A few moments later, he returned, and assured me that we could now safely continue the flight. We did just that. About half way to Atlanta, the flight attendant came to the cockpit and wanted to know what the first officer had said to the culprit. The man had not opened his mouth in an offensive way since we took off. He said, "I told him that the captain was the meanest SOB that he had ever seen, and it would not be advisable to agitate him further." I guess the man believed that I was who the copilot said I was. It worked. There comes a time when we must recognize the error of our ways and change our behavior. Please notice that I said, "change our behavior." I can possibly have an influence on the behavior of another person. But the individual, with the help of God, is the only one who can change his own attitude.

I heard a story, "The Vulgar Parrot," that illustrates this point. Fred received a parrot for his birthday. This parrot was fully grown with a bad manner and inferior vocabulary. Every other word was an expletive and those that weren't expletives were, to say the least, rude.

Fred worked hard to change the bird's attitude and continuously said respectful words, played soft music, and anything that came to mind. Nothing worked. He yelled at the bird, the bird got worse. He shook the bird, and the bird got madder and ruder.

Finally, in a moment of desperation, Fred placed the parrot in the freezer. For a few moments he heard the bird squawking, kicking, pecking, and screaming, and then suddenly all was quiet. The owner was worried that he might have actually hurt the bird and swiftly opened the freezer door. The parrot peacefully stepped out onto Fred's extended arm and said: "I'm sorry that I offended you with my language and actions. I ask for your forgiveness. I will try to check my behavior."

The owner was astounded at the bird's change in attitude. He was about to ask what changed him, when the parrot continued, "May I ask what the chicken did?" He had seen the frozen chicken while he was in the freezer. What does it take to shake us up to the point where we change our attitude for the better?

My family and I can relate to the story of this parrot. Our son, David, received a parakeet for his eighth birthday. The reason I am telling this story is to show the importance of focusing on the things that will help us control our attitude. *Thy word have I hid in mine heart, that I might not sin against thee. Psalm 119:11 (KJV)* We often called this our Christian bird. Jody, David, and Peter taught Freddie many different sayings. In addition to some general sayings, he was also taught to quote Scripture. Not only would he quote the verse, he also would quote the reference. Here are some of the examples:

> This is the day which the LORD hath made;
> we will rejoice and be glad in it.
> Psalm 118:24 (KJV)

> As for me and my house, we will serve the
> LORD.

Joshua 24:15b (KJV)

God is our strength.

Psalm 46:1 (KJV)

The joy of the Lord is your strength.

Nehemiah 8:10 (KJV)

Plus the following non-Scripture sayings:

Delta Air Lines, we love to fly and it shows.
Read the Bible!!
Western Air Lines, the only way to fly.
Mommy, rub my tummy, and he would tap the cage twice.

This may not seem like very much, but this parakeet had an incredible vocabulary. Let me tell you how this bird had an impact on those around him. When we would invite people over for dinner, Freddie would start quoting Scripture. One of the most often repeated remarks about him was, "That bird knows more Bible than I do." When he would start talking, he would not repeat himself. He would go through everything he knew in the order he was taught before he started over again.

This is significant in this instance. Jody and I were in the kitchen one day, and we were having an argument. I don't have any idea what the source of our disagreement was, but I do remember that Jody was having a tough time calming me down. About that time Freddie joined the conversation. He began saying, "Read the Bible, read the Bible, read the Bible." This is the only instance that I remember where he repeated himself. I can tell you that this was the end of our argument.

At another time, I was in the nook sitting at the table and for some reason I was a little depressed. Freddie was on my finger as I sat there feeling sorry for myself during my medical retirement. Out of the blue, the bird spoke up and said, "God is our strength, and it shows." This is another time that he got things mixed up, and I must say that it was appropriate for the situation. I knew that God was my strength, but it was not showing at that moment. God often gets our attention through some incredible circumstances. I have no doubt that it was God speaking to me through that parakeet.

Our treasured pet developed a tumor after only living a few years, and we had to have him put to sleep. Even in this dreadful situation, God had his hand on what happened during that week.

Our son, David was at a Christian camp during the week we put Freddie to sleep. He did not know this until he came home at the end of the week. While he was at camp, two of his friends gave their lives to the Lord. What was so significant about this was that both of these young lads had kept our bird in their home for a few days while we were on vacations. The verses that Jody, David, and Peter taught Freddie would not be idled with his death.

> So shall my word be that goeth forth out of my mouth: it shall not return unto me void, but it shall accomplish that which I please, and it shall prosper in the thing whereto I sent it.
>
> Isaiah 55:11 (KJV)

We need to remember that the best way to keep our attitudes pleasing to God is to continually focus on Him. We can do this in everything we do.

One of my copilots once asked me if it was alright to pray in the cockpit. I told him it was okay to pray at any time, any place. Before he asked this question, he told me about his grandmother who was gravely ill. We can have an attitude of prayer in any situation. We don't have to stop what we are doing, get on our knees, or bury our face in the ground to be in a prayerful attitude. We need to keep our focus on the One who gives us life.

How do we determine the importance of an individual? There are times when we put this label on a person for the wrong reasons. I read this little story in 2000+ Bible Illustrations on www.e-sword.net . I would label this man as important, because he clearly demonstrated his concern for his fellow man.

It Is Better to Be Late Than Pass by One Who Needs You.

Charles Harvey, good samaritan of Grand Prairie, Texas, felt a bit nervous as he drove to an important job interview. He was fifteen minutes late and hurried when he passed a middle-aged woman stranded with a flat tire. "My conscience made me stop," he said. "I changed her tire and headed to the interview thinking I could just forget about getting the job now." But he filled out the job application anyway, and went to the Personnel Director's office. Did he get the job? He sure did. The Personnel Director hired him on the spot. She was the woman whose tire he had just changed.

When we are out in the world, what kind of impression are we leaving on those with whom we come in contact? If we are members of the body of Christ, people should see the example of Christ in us. We need to be in the right Spirit, and we need to fully depend on Him to use us in a way that glorifies Him. As members of His family, we must live in a way that demonstrates the fact that we consider it a privilege to be part of His kingdom here on earth. In His sight, we are important to the One who is the giver of all that is great and good.

Boeing 757

I upgraded to this airplane, along with the Boeing 767, in October of 1990. This twin-engine, medium-range 757 is up to 80 percent more fuel efficient than the older 727 jetliners it was designed to replace, but retains the 727's short-field capability. The 757 can accommodate up to 228 passengers and has a range of approximately 3,900 nautical miles. The version that I flew carried just over 180 passengers with a normal crew of 5 flight attendants.

Taking Pride for a Ride

> Pride goes before destruction and a haughty spirit before a fall.
>
> Proverbs 16:18
>
> A man's pride will bring him low, but the humble in spirit will retain honor.
>
> Proverbs 29:23

Pride is often a feeling of superiority. It is sometimes manifested by a haughty attitude shown by people who suppose that they are better than others. A satisfaction with self, or the happy feeling somebody experiences when having or achieving something special that other people admire is another type of pride. One can feel especially pleased and satisfied to own or to have achieved something of importance. It is not too difficult to see that this term can have both positive and negative connotations. In my life, I have been guilty of using it both ways. Depending on what version of the Bible you use, this word appears around fifty times. It is presented in negative terms. So, is it okay for a Christian to exhibit pride? In order to answer that, I will present a couple of scenarios.

If pride is exhibited because of the first reason above, then I will definitely say that God hates pride. I believe it is okay to attribute success to God-given gifts. Woodworking is a hobby of mine, and I have been somewhat proud of many of the projects I have created. I even enjoy looking on some of these with a degree of delight. Anything I am able to do is because God gave me that ability. I participated by learning how to do some of those things, but it was He who gave me the desire and ability to do them. When taking all the credit, pride becomes negative. One may have reached the wonderful status of being a parent or grandparent. Being proud of the blessed arrival is within God's acceptable area of behavior.

Here is a story of a pilot in my squadron who was rather prideful. From the standpoint of the world, he had every reason to be proud. He achieved one of the highest scores you can get in the naval flight program. His grades were excellent in the areas of military, physical, and flight training. He was the student of the year. When he had accumulated enough hours to enter training as an aircraft commander, he almost breezed through the training. Before he was to take his final check ride, I was asked to be his instructor for the last training flight. This was a time when he could practice some of the maneuvers he would encounter on his final check.

I can tell you that it was a distinct pleasure to watch this man fly the C-130E. It was like watching and listening to a great musician play his instrument. I had never seen anyone who could fly with the precision that he displayed. Every maneuver he performed in my presence was executed flawlessly. There was only one problem: he knew he was good, and he didn't mind letting people know how good he was. Of course he was also good at taking all the credit for his abilities.

We completed the training flight and I told him what a joy it was for me to watch him fly the plane. I also cautioned him about his pride, and that it would one day rear its ugly head against him. He smiled and told me that I was entitled to my opinion. Little did I know that it would catch up with him so soon. I would doubt if this individual ever had a problem in any of his training.

After he returned from his check ride the following day, I was waiting for him so I could give him my congratulations. Upon greeting him, I discovered that he had failed to pass his examination. He failed to call for a critical item on an emergency checklist, which lead to his failing the evaluation. He had to go up the next day and repeat this part of the check, which he did successfully, thus becoming an aircraft commander. I reminded him that there would come a time in his life where he would view this as a valuable lesson. He could no longer claim that he was perfect, and he just may have swallowed some of his pride. God gave this person an incredible gift, but he thought it was because of his own efforts.

Another situation occurred involving pride with a student at Pensacola. While I was in training, there was a young Marine cadet who felt he was "God's" gift to aviation. He could prove this to you with the grades he had been getting. From the people I spoke with, he was quick to tell you about his accomplishments. He was another guy who could make the airplane sing. Again, his problem was that he was over confident. On the day he was qualifying on an aircraft carrier for the first time, dozens of new students beginning their training were on the carrier to observe the landings aboard the ship. After completing his eight required landings in an extraordinary manner, he decided he would provide a little entertainment for the observers on the carrier. As he was taking off the flight deck for his maneuver, he decided to do an aileron roll for the benefit of those watching. For some reason, when he rolled the T-28C upside down he never completed the roll. He flew into the Gulf of Mexico without completing his show. This was his final flight because he was killed in this tragic maneuver. You may have heard the term that "pride can kill." It was certainly true for this young man.

If you find yourself beaming with pride, just examine why you have those feelings. Are they present for the right reasons? God will enlighten you with the answer for which you ask.

> If any of you lack wisdom, let him ask of God, that giveth to all men liberally, and upbraideth not; and it shall be given him.
>
> James 1:5 (KJV)

If during this examination, you discover that your pride is the negative variety, what can you do to rid your life of this destructive tendency? God has promised us help if we will only go to Him with the right heart attitude. Sometimes we pray for improvement in certain areas of our life, but we still struggle to overcome some of those difficulties. Do we just give up, or do we keep on trying? We must work at the problem, with God's help, until it is resolved.

In June of 2000, I was flying with Steve. He had made what he thought was a decision for Christ when he was a teenager. During that time of his life, was turned on for the Lord. He was involved in all of the available youth activities of his church and enjoyed them. By the time I flew with him, he said that he had come to realize that all the things he believed as a teenager were untrue. He said he just couldn't make sense out of some of the messages he had heard by well-known preachers. I said I couldn't make sense out of some of those people either, but that did not have an effect on my belief in God. I simply advised him to go to the Bible and see if what the person was telling him was the truth. We had discussions for four days and I was unable to change his mind. He had taken a degree of pride in the fact that he possessed a knowledge that so many preachers fail to grasp. At this time, the Gospel no longer made any sense to him.

Though I quoted numerous Scripture verses during the course of the four days, I couldn't get through to him. I prayed for wisdom in order to say something that would make a difference. We taxied to the gate on the final leg of our four days. We prepared to shut down the engines, say goodbye to the passengers, and go home. The ground support personnel stood by to plug external power into the plane so that we could shut down the engines. I turned off the fuel to the engines, and as they wound down, the external power failed. Everything went black in the back of the cabin. The person on the ground reconnected the power cord, and after doing so, power was restored. I said to Steve that this experience was a lot like the Christian life. Sometimes we have to reconnect to the source that gives us power, which is often termed a rededication of our commitment to Christ. I encouraged him to connect once again to that wonderful source that had provided so much satisfaction for him as a young man. He promised to reevaluate his recent disbelief.

One day we will fly for the last time. Are you sure that you have the key to the right room when that time comes? After completing a long day of flying, I looked forward to checking into the hotel and getting a good night's rest. When we arrived at the hotel, a key would be waiting for us to enter the room for a period of rest and relaxation. There was a time when one of my copilots was given the key to the wrong room. He opened the door to the incorrect room, and a lady was taking a shower. I asked him later why it took him more than a couple of seconds to realize that it wasn't his room. He said he was tired and was not thinking quite straight.

If we have given our life to Christ, the key waiting for us to enter that heavenly room will be the right one the first time. We won't have to be concerned about entering the room of someone else, because Christ has gone ahead of us and prepared the right room for us. You can be assured that the key will permit you to enter the right room. Do you hold the correct key in your hand?

Boeing 767

I flew the last eleven years of my career in this type of airplane and the Boeing 757. Flight decks of the 757 and 767 are nearly identical, so when pilots are trained on either type they are qualified to fly both. The 767 was much larger. The 757 and the 767 were developed concurrently, so both shared the same technological advances in propulsion, aerodynamics, avionics, and materials.

The 767 is a wide-body, double-aisle jet, but, like the smaller standard-body 757, is designed for fuel efficiency. It can carry from two hundred to three hundred-plus passengers. The largest model that I flew carried just over two hundred fifty passengers.

Having the Right Credentials

Imagine that you are preparing to make your entry into Heaven. As you approach the gate, someone is checking credentials. Could you produce legitimate proof to permit your entry? Would you have the right identification, or would it be counterfeit?

When I was flying, one of the benefits of being a pilot for a major airline was our ability to fly on the jump seat of other airlines, as well as our own. This was a wonderful perk, especially if the flight we wanted to go on was full. All of that came to a screeching halt on September 11, 2001. I retired in July of that year, so I never had to operate under the increased restrictions. Now, airlines permit only their own employees to ride in the cockpit. Over the years, I met some rather interesting people by allowing them to use my jump seat.

I was in Seattle preparing for a flight to New York City, when I was approached by a pilot with a uniform that I did not recognize. I directed him to the gate agent, who was able to reference a computer listing to see if his employer was on the approved list. All of the airlines with this agreement were listed. The agent told me that we had a reciprocal agreement with his company for travel in the cockpit, but he needed to check to see if the pilot possessed a valid identification. At first everything seemed to be in order, but upon closer examination, the identification proved to be expired. He was therefore was not permitted to go on our flight. He had tried to make an unauthorized request for free travel. Had the agent been very busy, or distracted by other problems, he may have missed that important bit of information. In my experience, I know there have been times when people slipped by for various reasons. This could not happen today with the security checks now in place.

A few months later, I was in Atlanta when a man entered my cockpit and said he wanted to ride with us on the flight deck. He showed me all of his identification, but something was not quite right. He was not a pilot, but rather an employee for a company that makes aircraft engines. I escorted him out to the gate agent to see why he had been allowed to come to the cockpit. She had been distracted by another situation and assumed it was okay for him to fly with us, because he had arrived from another airport riding in the cockpit. He somehow circumvented the screening process and the gate agent made the assumption that he could sit in the cockpit since he had been approved earlier.

Had I not checked this out thoroughly, no problem would have occurred in this instance. He would have been approved to fly with us in the cockpit under certain circumstances. Since he worked for a company that made some of the engines we used, he could have taken steps to use this benefit. His company had to submit paperwork in advance that was approved by the vice president of flight operations. He could only fly in this capacity while performing required duties, which was not the case in this situation. He was trying to travel free for his vacation and it placed me in an awkward position to have to remove him.

Once when in Miami, I was approached by a gate agent who informed me that a pilot from another airline wanted to ride with us. He was with one of the top three airlines, so I knew that he met this part of the requirement. When he was asked to show the required identification, he could only produce two of the three that he needed. He said that he had left the identification in his flight bag on the other side of the airport. I told the agent that I could not accept him without this proof. She spoke to the individual and informed him of my decision. He became very angry with the agent. The agent politely informed him that my decision was final, and he left shouting all sorts of obscenities. He returned a few minutes later and said he wanted the name of the captain who refused his request, and he also said that my boss would hear of this. I gave the agent my business card, and told him I would be waiting for his report. If he had attempted to report this to my superiors, he would have placed his own job in jeopardy. This was a benefit that could only be enjoyed by following very strict guidelines.

In my early days of flying, we had kept the cockpit door locked while on the ground, and a person could only enter the front of the airplane if the right signal was given. On one occasion, I had a Federal Aviation Safety Inspector try to enter the cockpit by knocking on the door. He didn't ask the flight attendant to give the required signal, but he continued to knock several times to see if I would open the door. When he realized that he wasn't going to get in this way, he sounded the right code. He was then permitted to enter. As he came in the door, he said, "Congratulations, you passed." If we had opened the door without his knocking the required signal, we could have received a hefty fine. The entry signals in effect today are much more stringent.

People have attempted to gain unauthorized access for a long time, and I know many people who seem to think they can make it to Heaven on someone else's ticket. When it is time to go, a ticket must be presented that cannot be counterfeited. This is one event worth checking that the ticket is valid, because there is one chance to get in. You won't be able to say, "I left the real ticket in my other coat."

During my many years of flying, I carried a passport when traveling to international destinations. Some of the countries required a visa, an official endorsement authorizing the holder to enter, leave, or travel through a particular country or region. In August of 2000, I flew from Los Angeles to Mexico City. I carried four Korean passengers without visas. In Los Angeles, they boarded my airplane for the next leg of their trip. They did not have the required documentation, and I was forced to confiscate their passports. They were told that their passports would be returned to them by Mexican Customs and Immigration upon arrival. The agent who transferred these passengers mistakenly informed them that they would get them back once we were in the air. I found out about this error before departing and I requested that the agent return and explain the procedure correctly. This was necessary because none of the four passengers spoke English.

After departing, and during our climb, the Koreans decided that they didn't want to wait for their passports. They demanded that they be returned immediately. This presented the flight attendants with a bona fide problem because these passengers caused quite a disturbance. Fortunately, one of my attendants spoke broken Korean, and he was able to relay my instructions. I gave them the choice of landing in San Diego where they would be removed by the authorities, or we could continue to Mexico City. Then, and only then, would they regain their passports. They altered their behavior to my satisfaction, and we continued to our destination.

My job required the continual updating of credentials. As a captain, I had to have a physical examination every six months and my performance evaluated annually. Failure to do this would prohibit my ability to fly. The FAA often rode with us in the cockpit unannounced, and among other duties, made sure our credentials were up to date. The pilot's license, latest physical exam record, and a radio telephone operator's permit had to be in the possession of the pilot. The license had to show qualification to fly the airplane being flown. Each time a pilot checked out on a new aircraft, a new license was issued that showed all the airplanes he was qualified to fly. There were times when pilots could not produce one or all of these items, and they were not permitted to fly until they could prove they possessed them.

I began to carry a new set of credentials in 1969. This is when I became a Christian and was issued identification that has no expiration date. It cannot be stolen, lost, misplaced, or destroyed. One of these days I will use this, and I won't worry about being turned away for my inability to produce it. I received the stamp of approval that would assure my entry into Heaven when I signed on to be a member of the Kingdom of God. Are the credentials you have in your possession the correct ones? While people can be fooled by inappropriate credentials, God cannot be deceived. He knows who has His stamp of approval. A time will come when each of us must present our credentials to the One who makes the final decision.

Aircraft Carrier U.S.S. Lexington (CV-16)

This carrier was commissioned on February 17, 1943. The length of the ship is nine hundred ten feet, and the beam is one hundred forty-two feet. The vessel displaces thirty-three thousand tons of water and has a draft of thirty feet. All aircraft shown on the deck in this photo are North American T-28s. The Lexington is now a museum in Corpus Christi, Texas.

I had eight landings in a T-28C and eight landings in an S2F aboard this ship.

Unified in Purpose

> Behold, how good and how pleasant it is for brethren to dwell together in unity!
>
> Psalm 133:1

Flying is similar in many aspects to the church body. In the body, hopefully, we are proceeding toward the ultimate destination. Along the way, we sometimes get a little off course for a variety of reasons. The reasons are as varied as the number of people in the body. The parable of the sower in the thirteenth chapter of Matthew illustrates some of the major factors that heavily influence our lives, and we sometimes let things of the world distract us. A unified church body can help direct you back on course.

> Now he who received seed among the thorns is he who hears the word, and the cares of this world and the deceitfulness of riches choke the word, and he becomes unfruitful.
>
> Matthew 13:22

In flying, everyone who has a part in flight operations must be unified in purpose. The dispatcher has the responsibility for choosing the best possible flight plan. He must consider the weather, any operational limitations with the airplane, the duty time of the pilots and flight attendants, and any other influence that might have an impact on the safe operation of the flight. The maintenance personnel have the responsibility to make sure the airplane is mechanically sound and ready for the flight. The fuelers, caterers, baggage handlers, ground controllers, and tower operators are some of the other people involved. All of them must have their eyes on the same goal to get the passengers to their desired destination in a timely and safe manner. There must be unity.

A story that illustrates the importance of being of the same mind and purpose took place in the late 1970s in Sacramento, California. A short time before this flight, the FAA issued a directive. It required the pilot flying the airplane to brief the pilot who is not flying on the details of the standard instrument departure procedure for that airport. This chart gave clear instructions of what to do after takeoff. It was composed of heading and altitude directions and restrictions. Failure to comply with the printed instructions placed the pilot and passengers in jeopardy. Please understand that this was a new requirement and some pilots struggled to adjust to the new procedure. Before this directive came out, we actually had the same requirement, but it wasn't spelled out in a hard and fast rule. This new technique came about because some failed to do what they should have done all along.

As we taxied out the departure runway, the flight engineer came to the part on the checklist where it called for departure briefing. My copilot answered that it was complete, because he was going to be the pilot flying on this leg. He didn't like the FAA telling him how to fly the airplane. I then asked him to humor me by telling me what his intentions were after we lifted off the runway. This particular departure required very specific actions shortly after becoming airborne. Without referring to the printed chart, he proceeded to tell me about the heading to which he would be turning and the altitude to which we would climb. After he finished, I calmly reminded him that I could not allow him to do that. We stopped the taxi, and I requested that he take a close look at the chart. It didn't take very long for his face to turn beet red. He quickly saw that we would have been handed a violation for turning into a high density traffic area. We became unified in purpose, and continued the flight in a safe and timely manner.

I understood why he gave the erroneous briefing. When we landed in Sacramento, we made our approach and landing to the South. While we were on the ground, the wind changed, and we made our departure to the North. If we had taken off to the South, the instructions he gave would have been correct. He simply failed to consider the runway change. I think he quickly understood the importance of why this procedure became mandatory. Checklists are very clear about what is required so the pilots have no misunderstanding. The checklists used today are the best they have ever been. If we used similar checklists in our churches, we might become a little more unified. You wouldn't have as many who insisted on doing things their own way.

Have you ever had the opportunity to watch one of the military precision flying teams such as the Navy Blue Angels or the Air Force Thunderbirds? The pilots on these teams must be unified. When they are in the middle of one of their multiple plane formation maneuvers, one of the pilots cannot decide he is going to do something a little differently. They must perform those fantastic precision aerobatics in concert and harmony. Wouldn't it be great if we could learn some lessons from these types of teams? Life is much sweeter if we walk in love, unity, and harmony with our fellow man.

When I was in the flight training program at Pensacola, I had a certain amount of formation training. We were not trained in the types of activities such as the formation aerobatics that the Blue Angels and others perform, but some of the training was the same. In a formation with other airplanes, you placed your trust in every other person in that formation. You must be united in purpose, and the leader must be followed. However, if he fails in his responsibility to lead in the proper manner, he sets others up for disaster. I remember a time when I was tucked in close behind the leader of my formation. In one particular instance, he decided to slow down without giving the proper signal to the rest of the formation. At this time, I was in the number two position when he extended his speed brakes and began a rapid speed reduction. Without knowing what his intentions were, I did not have time to respond, and passed him like greased lightening, coming dangerously close to his aircraft. To this day, I'm not really sure how I missed him. I was not a Christian at this time, but I believe I said, "Thank you, Lord." For a moment, the leader and I were not unified in our intentions.

Many years ago, I was returning from another trip to Vietnam. I was copilot for a Navy commander. In the early days of flying, the aircraft commander and captain were almost gods. If they gave you an order, you were to do it without question. This attitude resulted in disaster on more than one occasion. There were a number of times when I did question the reason for taking certain courses of action. The military is certainly not perfect today, but the operational procedures have changed and there is a greater degree of accountability.

On one segment of the flight from Vietnam, we were going to fly from Wake Island in the Pacific to Travis Air Force Base in California, which would take over fourteen hours. We were flying an empty airplane which permitted us to carry considerably more fuel. After we were in the air for a little over two hours, we had an engine failure. This is the third time the same engine had failed on three consecutive flights. The engine had been changed on Mactan Island in the Philippines, where we stopped on some of the trips to Vietnam. We had a number of mechanics working on the airplane who had never changed an engine on a C-130E. It took over twenty-four hours to complete the work, performed with a wrench in one hand and an instruction sheet in the other. The mechanics were dedicated people, but in this instance, they didn't have enough experience. The first time the engine quit was when a hydraulic line broke loose and dumped all of the hydraulic fluid into the engine. This resulted in the engine shutting itself down without damage. The second time, an oil line came apart and caused the engine to quit. Both situations happened because the lines had not been properly tightened.

When the third engine failure occurred, we knew we would not make it to Travis Air Force Base. We had to consider our options. The aircraft commander said we were going to Midway Island because it was closest, and he ordered the flight engineer to dump most of our fuel. By dumping fuel we could remain at a higher altitude and go a little faster, but we would now have only one choice of a place to land. This is where I had a problem with the decision. I suggested that we check the weather on Midway Island before we dumped that much fuel. If the weather were a problem, we could make it to Hawaii as an alternate solution. In no uncertain terms, he informed me that he was the aircraft commander, and we would abide by his decision. I was accused of talking when I should have been listening and obeying.

The fuel was dumped as ordered, and we were committed to land at Midway. As we approached the island we checked the weather and discovered that it was not very good. The winds were so strong that we could not land on the longer runway, and the crosswinds exceeded the limits for the plane. We had to land on the shorter runway, which is no problem for a C-130, but you have to be able to see the landing strip. This was before airplanes had the capability of doing a fully automatic landing. By the grace of God, we were able to land safely. I want to tell you that the approach was full of drama. This could have been avoided if the man in charge had listened to the members of his crew. He was a majority of one with his decision. This pilot lost his aircraft commander certification a short time later. He had outlived his usefulness in his position. He became inflexible because no pilot who was going to tell him how to perform his job.

Emergency procedures are designed today to include both pilots in the cockpit. Before any important switch or lever is actuated in an emergency, the other pilot must verify that you are doing what you are supposed to do. This promotes a more unified cockpit. In this type of environment, all the cockpit crew members have a sense that they are important.

Pictures with Stories

This picture was taken in the cockpit toward the end of my career at the Miami International Airport before departing for Cincinnati, Ohio. We spent the night at a hotel, and arrived at the airport before noon for a flight that was supposed to take us to Cincinnati and then on to Salt Lake City. It didn't quite turn out that way.

Archie in cockpit of Boeing 767 February 2001

While my flying partner conducted his preflight inspection of the plane, he found a small fuel leak under the engine on the right wing. We called the mechanics, and upon examination of the small leak, they said that an insignificant amount of residual fuel had spilled. In their opinion, it was safe to proceed. I told them that it wasn't quite that easy because I had made an entry in the mechanical logbook. This entry required that they conduct a closer check of the problem, which included an engine start at the gate to see if the leak was of any consequence. With the mechanics positioned outside the airplane, I started the engine. As soon as it started, they signaled me to shut it down. Fuel was now coming out at a rather fast rate. Upon checking inside the engine at the source of the leak, they discovered a small crack in one of the fuel lines which now had to be replaced.

Mechanics hard at work repairing fuel line

Our Christian walk is a lot like this fuel line. If something is not quite right, we have a tendency to think that if we leave it alone it will take care of itself. When these things happen, it may be because we are receiving warning signals to indicate the possibility of a more serious problem. We should not be so quick to discount these warnings. It is certainly easier to rectify small things in their early stages. Nipping problems in the bud before they are fully in bloom will enhance your Christian witness.

We were scheduled to fly to Cincinnati with a full passenger load. The part we needed was not in stock in Miami, and we waited over three hours for the part to be flown in from Atlanta. Four hours later, the part arrived and the engine was repaired. We were once again ready to depart. The gate agent closed the door, and we pushed out from the gate and began our taxi toward the active runway. Our original passenger load was now at two-thirds capacity, since the other one-third found seats on other flights to arrive at their destination a little sooner. The copilot read the taxi checklist as we taxied toward the runway. While he read and I answered, I noticed that one of the cockpit gauges indicated a significant loss of hydraulic fluid. This meant a return to the gate to check out this mechanical abnormality.

The mechanics met us at the gate and immediately noticed that we were losing hydraulic fluid near one of the wheel well areas. One of the brake lines had cracked, causing us to lose the fluid. We were faced now with another wait to get the line fixed and after another four hours, the plane was certified to continue. All of our passengers had made other arrangements for travel to Cincinnati, and we knew we were not going to arrive home in Salt Lake City that night. We flew an empty airplane to Ohio and spent the night there. This was a situation where you have several creeping delays. At first the problem seems to be one that could be repaired quickly, but it doesn't always work out that way. Schedules are important, but safety is always the first consideration. I can also tell you that pilots don't like creeping delays any more than passengers do, but we don't want to be on an airplane that is not completely safe.

During my last few months prior to my retirement in July of 2001, Jody traveled with me about one trip per month. We had the opportunity to visit great places and spend some time together. She accompanied me on a trip to Norfolk, Virginia, where we were able to spend time there and see longtime friends. One such friend, Stan Jeter, who was the best man in our wedding in 1972, lives in Virginia Beach with his wife Rhoda and their family. It was a wonderful break from the normal routine of an ordinary layover. Jody also went with me to New York City where I had a double layover, with a trip to Nassau in between. The first layover was for over twenty-four hours, and we had time to do some great sightseeing. My copilot and I flew a charter flight down to Nassau and back, while Jody remained in New York to tour the city. This was also a great trip for me, because I had Jody with me and I had never been to Nassau. I was getting to spend terrific time with Jody as well as seeing a new part of the world.

Jody sitting in my lap before takeoff

On the charter flight to Nassau, we took a plane load of employees from CITI Group down to Nassau, and we were supposed to fly an empty plane back to New York. When we arrived in Nassau, we were told that we were going to have six passengers going back to New York with us. We also had six flight attendants on board. As we departed the airport and climbed to a safe altitude, I welcomed the six passengers on board. At that time, American Air Lines was running ads touting the generous amount of legroom they offered on their flights. I told the passengers about that, but said we had something more special here at Delta: "We have one flight attendant per passenger." Everyone got a kick out of that little comment.

I always tried to let my passengers know how much I appreciated the opportunity we had to serve them.

This next photograph was taken during my military days in March 1967. This was my last flight in a C-130E Hercules. You will notice that it has U.S. Air Force markings. I was in one of three Navy squadrons, VR-22, VR-7, and VR-3 that were flying Air Force planes, under the umbrella of the Military Airlift Command. The only thing that showed we were a Navy crew was the abbreviation on the side of the airplane, near the front, NATWP. This stood for Naval Air Transport Wing Pacific. A pilot in my squadron took this picture of me as I flew above the clouds over San Francisco.

Archie flying C-130E over SFO in March 1967

I flew my thirty missions to Vietnam in this type of aircraft. In three years of flying the C-130E, I logged over two thousand hours of flight time. I finished my career with over twenty thousand hours of time in the air that included both flying for the military and commercial flights.

In the next photo I was standing beside the engine of a Boeing 757. This particular airplane was painted with the Salt Lake City 2002 Olympic paint scheme. The 757 is a beautiful airplane anyway, but it especially caught people's attention adorned in these brilliant colors. When this picture was taken, I had just arrived from Anchorage, and I was now getting ready to go to Cincinnati.

Archie by left engine of Boeing 757 in SLC

Boeing 757 cockpit

When Problems Arise

The event occurred early in the 1970s. I was flying primarily from San Francisco to Honolulu, returning from a wonderful layover on the beautiful island of Oahu. I was the flight engineer on this flight, and it was my responsibility to do the preflight inspection. While I was in the process of doing the walk around portion of my pre-flight inspection, I discovered a small piece of metal in one of the view ports of one of the four engines of the Boeing 720 aircraft. This is essentially a small Boeing 707 and, except for the size, it is difficult to tell them apart.

Boeing 720B on takeoff from San Diego

Upon finding the metal, I called for a mechanic to examine the problem further. After he checked everything out, he pronounced the airplane safe to fly. I didn't agree with his opinion, so I contacted the captain and told him about the problem. He was inclined to agree with me, and he didn't want to take this airplane on a flight to San Francisco without a more thorough check. After a lengthy consultation, we agreed to take the plane out to a remote part of the airport and run the engine up to full power to see if everything was fine. On the third attempt, we felt a vibration coming from the suspect engine. That was sufficient for the captain to cancel the trip. We went back to the hotel and waited for another plane.

When we left for the hotel, the mechanics did further tests on the engine and could not find any significant reason for grounding the ship. The company called another crew to fly it back to the mainland. After taking off from Honolulu, the engine failed and caught on fire. The crew returned safely to the airport with no injury or further damage to the plane. This time, the mechanics removed the faulty engine and found a crack in its side. They also confirmed that the metal I originally saw was from this crack.

This did not happen because of poor mechanics. At Western and Delta, I believe we had the best available airline mechanics. The maintenance personnel followed the procedures correctly, but there was something they couldn't see in their normal inspection. After this incident, additional checks were added to their procedures. I was involved in a few situations where procedures had to be changed because of new discoveries.

When you discover something is wrong, make sure you thoroughly satisfy yourself with the fact that the problem is solved before you continue. As Christians, we need to continually examine ourselves to see if we are living our life in a way that brings glory to Jesus Christ.

In another instance, I didn't give the mechanics as much credit because they simply didn't evaluate the problem sufficiently. We had landed in San Francisco after a flight from Seattle and, before we continued to our next destination, I performed an enroute walk around examination to make sure no new problem had developed. Problems could arise, for example, if we had hit a bird in flight or if the plane had developed a leak in the fuel or hydraulic system. We did our best to assure the safety of every flight. On this inspection, I found a hydraulic leak coming from one of the main landing gears. I showed the leak to the mechanic and informed him that I was going to make an entry in the airplane's maintenance log. He said he was just going to wipe off the fluid and sign it off, thus dismissing the opportunity to inspect the gear for potential problems. This was not acceptable to me, so I called for his supervisor and had him take a look. After he saw the problem, the airplane was towed to the hanger and repaired. When we have a problem, we can't push it under a cover and ignore it. We need to correct that situation before it becomes a greater problem. The captain thanked me for my persistence.

Except for my fourteen year close call, this next story is near the top of the list of the closet calls I have ever encountered. In 1971, we were scheduled to fly from Los Angeles to Acapulco with a full load of passengers and a maximum fuel load. This was because the fuel was cheaper in Los Angeles than it was in Mexico at that time. This was to be a trip that was different from anything I had ever experienced up to that time. In the remaining thirty-plus years of my career, I never had another trip that generated this much excitement.

When we left the hotel, I had an uneasy feeling about the flight. I didn't have any clue as to why I felt that way. We made our normal preparations for our flight, and we were now ready to depart. We pushed back from the gate, started our engines, and began our taxi out to the runway. When we were underway on our own power, the copilot's compass failed. We returned to the gate and that problem was corrected. Once again we began our taxi, but this time one of our tires went flat, and we had to return to the gate. I was beginning to understand why I had an uneasy feeling about the flight. Both of these situations were highly unusual. This was the first and only time I ever had this kind of uneasy feeling before I worked on a flight.

We taxied out once again and this time made it to the runway. The weather in Los Angeles was rainy with a ceiling of about two hundred feet, which meant that the base of the clouds was two hundred feet above the ground. The runway was wet and the wind came from the east, the direction in which we would take off. After a few minutes, we were cleared for our takeoff on runway seven. When we takeoff, we are constantly aware of matters related to safety. We have a number of airspeeds with which we must be concerned, two in particular being the V_1 and V_R speeds. The first is defined as the critical engine failure speed. If a significant problem arises before reaching this speed, the takeoff must be aborted and the problem handled on the ground. If this speed is passed and a serious problem develops, the takeoff is continued and the emergency, or abnormality, is handled in the air. The second speed is the speed at which we start our rotation to become airborne.

Everything appeared normal as we passed the first speed. When we reached the rotation speed, the captain tried to start his rotation and was unable to do so. The controls of the airplane were locked up and could not be moved. We were almost twenty knots over the rotation speed when the captain aborted the takeoff. In stopping the aircraft, we blew a few tires and stopped just a few feet before the end of the runway. If he had delayed the abort another fraction of a second, we would not have been able to stop. Once we passed the first speed, we were, in theory, beyond the point of being able to stop the plane on the runway if a problem arose. We just happened to be on a very long runway. Some of the runways that we fly into are much shorter than the one in Los Angeles, but this is one time where we benefited by having the extra length.

So why did the controls freeze? When we picked up the airplane at the gate, it had just come from the hanger where it had undergone maintenance on the elevator system. An elevator bell crank was replaced, and it was installed upside down. This is not something the pilot can find on his inspection. The bell crank looks the same on both sides and appears that it would work this way. A closer look reveals a difference. Three mechanics signed off on this maintenance action after completing the change, and they were still wrong. But for the grace of God, this nearly resulted in disaster, and was another situation that caused Boeing to make changes to their inspection procedures. This was the first time this had happened. Since drastic measures were undertaken by Boeing after this incident, this type of problem would never happen again.

Norton Simon, founder of Norton Simon Inc., a conglomerate of consumer product companies, was a passenger on that flight with his new wife, Jennifer Jones. She was an actress who had a supporting role in the movie, ***The Towering Inferno***. Simon amassed a huge fortune that he invested in one of the world's greatest collection of artwork. Those pieces of art are housed today in the Norton Simon Museum in Pasadena, California. Mr. Simon passed away in 1993, but he came close to departing this life a little early on this flight to Acapulco.

About a year before I retired, I flew to Mexico City. After completing my layover, I prepared for a flight back to Atlanta. When I arrived at the airplane, I was met by a maintenance inspector from Atlanta. The airplane we were to fly had an engine failure upon landing, and the engine had to be replaced. The inspector had supervised the engine change and had since discovered a small hydraulic leak in the engine. He wanted to know if I would fly the plane back to Atlanta with this small leak. I left the confines of the cockpit to examine the leak coming from a line that led to one of the components on the wing. I asked the supervisor if they had started the engine and put pressure on this line to see if the leak worsened. He said they had not, but that it was good idea, and proceeded to start the engine. The fluid streamed out at a rapid pace. The flight was not going to happen until the problem was resolved. I suggested they check to see if the connector on the line had been properly tightened when the engine was changed, and they found the nut to be loose. The problem was solved by a little tightening, and we departed for Atlanta without further incident.

When our son David was about two or three years old, we were riding in the car and he was strapped in the car seat in the back seat. It was dusk and the moon was out. Suddenly he exclaimed, "It's twixed, it's twixed, it's twixed." His enthusiasm filled the car. He was clearly excited. Jody and I were puzzled by the statement, and we asked him what he meant. He said, "The moon, the moon, it's twixed." We then realized what he was saying. A few nights before, he had noticed a partial moon. It was now full and no longer needed to be fixed. This is the way it should be with us. When something in our lives is broken, there should be joy when it is fixed. As a friend of ours once said, "Joy is not the absence of suffering, but the presence of the Lord." What a joy it is to be in God's will and be "fixed."

The Choices We Make

Life is full of opportunities to make choices. When we get up in the morning we have to decide what to wear, what to eat, whether to exercise, and so forth. Some choices are swayed by external influences. When we leave for work, the day will present many situations which demand all sorts of decisions and choices. The choices we make have everything to do with the peace and contentment, or the lack of it. The choices we make can be seen in a little different light. Here are a few illustrations about choices certain people have made.

In the later part of the 1970s, I was flying from Seattle to Los Angeles. Half way through the flight, one of the flight attendants came to the cockpit to announce that she had a female passenger who was causing problems for the crew. A lady had consumed too much alcohol, and she was intoxicated. While under the influence of this substance, she went to all the doors on the airplane and tried to open them and get out. Little did she know that this is next to impossible, because of the pressure per square inch that is placed on the door with the pressurization of the plane. The difficulty did not stop her from trying, and she could not be controlled by any of the cabin crew members.

I left the cockpit with the copilot in charge of flying the airplane, and went back to converse with the lady. I found her in the first class galley after one of her attempts to open a door. I asked her what her problem was, and she said, "Sir, I don't have a problem!" I said that I would like to keep it that way, so I gently escorted her back to her seat. By now we were about forty minutes from landing. In our discussion at her seat, I asked for her cooperation by requesting that she remain in her seat for the rest of the flight. I outlined the consequences for failure to comply. If she caused any more problems for the flight crew, I would call the authorities in Los Angeles and have her escorted to jail upon arrival. She assured me that there would be no further problems, and then she told me that none of this was her fault. I couldn't wait to hear her explanation of that statement. She proceeded to tell me that all this happened because the flight attendants had been pouring liquor down her ever since she boarded the airplane. I responded by saying that you really have to watch out for those flight attendants! They will do that every time. After a few minutes with her, I returned to my duties in the cockpit. She caused no further problems for the crew.

When I presented this lady with the consequences for failure to comply with my instructions, she had a choice to make. She could continue to Los Angeles without causing further trouble, or she could resume her actions and spend the night in jail. She did not have to think very long about what she was going to do. If we were faced with black and white consequences before we made our decisions, we might be more careful in the choices we make. In my volunteer financial counseling sessions, I encourage people to think of the possible results of the decisions they make. I do this primarily in the area of handling money, but the principles work in other areas as well.

When you have a decision facing you, make use of the PTA concept that I have been teaching for years. Pray about the decision, think about the consequences, and act upon the decision when you have the peace God gives you.

Here are some steps you might consider when you have an important decision to make:

- Pray about the decision. (PRAY)
- Think about the decision. (THINK)
    - Define the dilemma.
    - List the alternatives such as the pros and cons.
    - State the criteria.

    - Evaluate the alternatives of the choice available.
- Now you are ready to act. (Make your choice.) (ACT) "Remember to Pray, Think, and Act (PTA)."

Sometimes we reap the effects of choices others make in regard to things concerning us. Here is a story that includes choices made by several different people.

I was in the last six months of military service when my boss, a Navy Commander, came to me and asked me to do him a favor on a training flight. A pilot in our squadron, a lieutenant commander, had completed over sixteen years of military service. He was transferred to our unit from another type of flying where he had accumulated over ten thousand hours of flight time. My boss asked me to take him up on a training flight and recommend him for aircraft commander training. One thousand five hundred total flying hours in all types of airplanes, and over one hundred hours in this type of plane are necessary for this training. He had greatly exceeded these requirements, but lacked one important thing. He had not been recommended for upgrade training by any of the pilots he had flown with in over one full year. I was asked to give him that approval.

I took the pilot on a four hour training flight where I observed many of his techniques and abilities. He did a wonderful job flying the airplane, but problems surfaced when we conducted approaches and landings. He executed the required approaches in a satisfactory manner, but had a problem with his landings. I had to take over and complete every landing that we made. I could not determine the cause of his inability to land. He appeared to have lost an important part of his ability to fly.

I returned to my base and went to see my boss to give him my report. I told him that I could not give a recommendation for upgrade training. In fact, I said that I would advise that he not be allowed to fly even as a copilot.

This gentleman had an unblemished military flying history and no known explanation for what happened when we flew together. My boss said that he was going to override my recommendation and train the pilot as an aircraft commander. I told him that this was his choice, but I would not change my report.

The next day, the pilot began his training with another instructor. On their first approach and landing, they lost control of the airplane on the ground, exited the runway, and hit a small structure that contained electronic equipment for the airport. This accident resulted in one of the engines becoming separated from the plane, as well as other considerable damage. The upgrade training for this pilot was terminated. When the new instructor discovered that I had flown with this man the day before, he was infuriated because the nature of my report was never revealed to him. Flying is a job that requires one hundred percent vigilance. When a pilot performs most of the aspects of his job satisfactorily, there is a tendency to become complacent at other times. Even though we don't like to think about it, we must never lose sight of the fact that anyone can make a mistake. The job of the non-flying pilot is to monitor the performance of the pilot flying, and he must be ready to assist when absolutely necessary.

Why did this have to happen? We must listen to any warnings that are available to us, and we must carefully evaluate the nature of those warnings to determine if they will have a detrimental effect on our plans. The signs are there much of the time, but we must comply with what they are telling us. Make your decisions in a manner that will produce positive effects on you and others.

Earlier I told you that I was flying near Dallas when President John F. Kennedy was assassinated. I was also in Boston, about fifty miles north of Martha's Vineyard, in July of 1999, on the day that John F. Kennedy Jr. was killed in a plane crash. As we approached the runway, the tower reported clear skies with reduced visibility. It was extremely hazy. We did not see the airport until we were below one thousand feet altitude. This was not a concern for us because we often executed approaches in far worse conditions with the aid of our sophisticated equipment. When we finally saw the active runway, I told the copilot that I was glad I was not flying a single engine airplane with the hazy conditions that existed. Late that evening I learned about the crash of the only son of the late president. He was going to a wedding, but departed New York City later than he intended. This meant that they would be flying after dark, and flying over water in poor weather conditions. This is a hazardous combination for an inexperienced pilot with limited instrument training. In good weather, you can see the lights on the horizon for an attitude reference. This is not possible when the lights are hidden by the haze. If you have an attitude reference (horizon), it is easier for an inexperienced pilot to keep his wings level. Without that reference, it is very common to become disoriented. Because I was in Boston earlier that day, and was familiar with the meteorological conditions that existed, I believe John F. Kennedy Jr. became confused and lost control of his airplane, resulting in the death of all on board.

> Your word is a lamp to my feet and a light to my path.
>
> Psalm 119:105

I firmly believe that young Kennedy knew some of the risks he faced when he departed New York City that evening. Early in flight training, stern warnings are given about exceeding your limitations. When he realized the lateness of the hour, he had a choice to make. He could either fly his airplane or to drive. He should have considered a number of things. He had broken his ankle in a sports game a few weeks before and his foot was still bandaged. Someone who saw him at the airfield said he was carrying a crutch as he limped to his plane. The rudder on his Piper Saratoga II aircraft is controlled by the feet. In addition to all this, he was not qualified to fly on aircraft instruments alone. He was only permitted to fly on "visual flight rules," which is when conditions allow a pilot to see where he is going without the aid of instruments. He made the choice to fly because it would have taken him over seven hours to drive there. This is a classic textbook example of someone taking the easy way out when it was not the prudent thing to do. One of my copilots told me an interesting story that illustrates this point very well.

The copilot and her husband flew to England for a vacation, and they decided to fly the Concorde back to Washington, D.C. After they left the gate in London, the pilots were told to bring the plane to a remote part of the airport because of a report that a bomb was on board. Everyone left the supersonic aircraft while it was thoroughly searched for explosives. After four and one half hours they were cleared to get back on the plane. When the passengers were all back on board, the captain made the following announcement. "Ladies and gentlemen, I would like to welcome you back on board, but I am not going to apologize for the delay. It is better to be four and one half hours late to our destination than to arrive early in the hereafter." There is a considerable amount of wisdom in that statement, and it would behoove others to apply this philosophy when faced with choices.

During the Vietnam War, I made a decision to disobey the rules. The only time I would ever consider doing such a thing was when safety was a prime concern. My crew and I landed at an airbase in South Vietnam where an Alaska Airlines C-130 waited to takeoff at the end of the runway. About an hour and a half later the plane was still sitting in the same location with the engines running. I asked the control tower why the plane had been in that position for such a long time, and I was told that the pilot was waiting for air traffic control clearance, but that there was a problem with communications. The procedure mandated that we must have an instrument flight rule (IFR) clearance before departing. After my crew finished removing the cargo and were ready to depart, we were also unable to contact the agency who issued the clearance back to Okinawa. I had a decision to make and the key reason that helped me to decide was the fact that heavy fighting was taking place around the perimeter of the airport. I could see the bullets stirring up the dust, and I wanted to get my crew out of there as quickly as possible. After conferring with my fellow workers, I decided to depart without a clearance. My plan was to obtain authorization once airborne. When we taxied out and took the runway, the Alaska Airlines plane was still waiting at the end of the runway. We finally received the required approval about an hour after takeoff. We safely departed an area of heavy conflict, and there was never a question about my judgment regarding this matter. I had a valid reason for deviating from established rules because of the safety issue.

When faced with a decision, do you take advantage of all of the resources available to you? We have the Word of God as our resource, but we must know what it says and act accordingly. I hope and pray that you will see the error of your ways if you try to make things happen your way. May God grant you the wisdom to make the right choices.

> There is a way that seems right to a man, but
> its end is the way of death.
>
> Proverbs 14:12

The Past Meets the Present

Nearing the end of my career, I arrived at the gate in Salt Lake City to board the airplane where I would conduct my preflight duties. I had not met my crew, including my copilot, for this four day trip. When I signed in on the computer at the chief pilot's office, I knew the names of those with whom I would fly. The last name of the pilot joining me on this trip seemed familiar. He arrived a few minutes after me, and we introduced ourselves. His name was Bill Records. I mentioned to him that I had flown with someone over twenty years ago named Bob Records, and then learned that Bob was Bill's father. Upon making this discovery, I fondly recalled what a delight it was to fly with his dad.

On the first day of our flight, we flew a total of four flight segments, or legs. After we settled in at the hotel in Phoenix, Bill telephoned his father in northern California and told him that he was flying with me. The son asked the father if he remembered flying with me all those years ago. Bob said that he not only recalled the times we flew together, but he was looking at a Bible that I had given him on one of those flights.

We flew through Salt Lake City on our way to Phoenix, where Jody joined us and flew with us for the next three days. In Phoenix, Jody and I had breakfast with friends. The next day we flew to Norfolk, Virginia, and had lunch with friends we've known for over thirty years.

Our time of visitation came to an end, and we were ready to depart for Atlanta, and then on to Salt Lake City. When Jody and I travel with airline privileges, we are on a standby list. This simply means we get to fly for free if there is a seat available on the flight. For my entire trip Jody had been very fortunate to get a seat on each of my flights. As we walked through the concourse on the way to the airplane, Bill made the statement to Jody, "You have been very lucky to make all of Archie's flights." Jody replied, "Bill, I don't believe in luck." Bill said, "What do you call it?" Jody responded by saying, "I like to think of it as God's blessings and God's challenges." Bill didn't quite know how to respond to this statement. This had some significance a little later.

At the gate, we discovered that the flight to Atlanta was oversold and that Jody would not make it to Atlanta on this flight. We said our goodbyes and said we would see each other back in Salt Lake City. I was now scheduled to arrive about two or three hours ahead of her, and Bill offered to give me a ride home so that Jody would have the car at the airport. Jody planned to catch the next flight out of Norfolk about an hour after we left, which caused her to have to wait longer for her connecting flight home in Atlanta.

Bill and I flew to Atlanta with a full passenger load, and after about an hour in Atlanta, we departed for Utah. We were level at our cruising altitude in the vicinity of Memphis when Bill asked me if I smelled electrical smoke. My ability to smell is not the greatest and I did not smell anything. Because I know my limitations with my sense of smell, I would not permit myself to rely on my ability. I called to the back of the airplane and asked my flight attendants if they were cooking something in the first class cabin, and they were not. I then invited them to the cockpit to see if they smelled anything unusual. When they entered the cockpit, they confirmed the smell of electrical smoke. Upon hearing what they had to say, I declared an emergency and headed the plane toward Memphis. We were never in danger, but discovered after landing that the equipment cooling fan had burned up without giving us the normal cockpit warning. This fan is used to cool the equipment related to our flight instruments, and without it, we would have eventually lost the use of our critical instruments. If we had delayed our actions for a considerable amount of time, the situation could have become more problematic.

We also learned that there was no replacement cooling fan in Memphis, and that the part would be flown in the next day. Jody was going to beat me home, and I did not have any way to get in touch with her. She arrived home at about two thirty in the morning, wondering why I hadn't left a light on for her. She discovered why I hadn't turned on the lights when she reached our bedroom. I had never made it home.

Jody called Delta immediately. They informed her that I was in Memphis and that my trip had changed. They did not tell her what had happened to my flight. It turned out to be a blessing that Jody and I were separated in our travel. I had many responsibilities with the maintenance crew, as well as seeing to the care of the passengers. They were not going to make it to Salt Lake City that evening.

My crew and I departed for the hotel in Memphis. We were scheduled to come back to the airport the next morning to continue the flight after the problem was repaired. We did not fly directly to Utah, but instead we flew an empty plane to Dallas, and then we continued to Salt Lake City as passengers. Instead of getting home at around 10 p.m. on Friday, I arrived after 2 p.m. on Saturday.

As Bill and I flew the empty 757 to Dallas, we picked up our conversation from the previous day. I said, "Do you remember Jody's reply to you when you told her she was lucky?" He said that he remembered. I told him that Jody and I would label the episode of the previous day as one of God's blessings. Bill and I had encountered one of His challenges.

I had an opportunity to present the Gospel to Bill during our time together, and I gave him a copy of the New Testament. This is the only time that I had given both a father and his son a copy of the Bible. I had the opportunity to fly with four or five father and son combinations in my career. I flew as a copilot for the fathers, and their sons flew as my copilots. I always enjoyed these unique experiences.

Times of Sharing

The Christian rapture has been described in many different ways over the years. Some have ventured to call it a supernatural transportation, as it relates to the experience of being transported into the spiritual realm. It is most often applied to the second coming of Jesus Christ, when His true believers are expected to be gathered together to join Him in Heaven. In the past few years a number of books have been written about this subject, primarily the "Left Behind" series by Tim LaHaye and Jerry Jenkins. Many people have come to know Jesus Christ through exposure to these books and their message. I have often said that I hope the rapture takes place in my lifetime.

In the late 1970s, I had my first opportunity to fly a series of flights with a cockpit crew who were all Christian believers. My copilot was Charlie Coxe, a dedicated believer who kindheartedly promoted the Kingdom of our Lord and Savior. He later contracted cancer and is now at home with our Lord. My flight engineer was Ron Montoya who was also a committed Christian.

I have had wonderful times of sharing on the flight deck with my fellow crewmembers, but I can't remember a more intimate time than I had with Charlie and Ron. The Lord was never far removed from any of our conversations, both in and out of the cockpit. Flying with these guys was like having a preview of what it is going to be like in Heaven. When we arrive and sit down in our Heavenly chat room, our focus will never be divided. Our praise and focus will continually be on the One who made it possible for us to be there.

During one of our many enroute stops, we had a few moments to relax before proceeding to our next destination. I was flying the Boeing 737 at the time, and we had three flight attendants aboard, two of whom were born-again Christians. They came to me and told me that they had been sharing the Gospel with their other crew member without any success, and asked me to talk to her. Without any hesitation, I left their company and sought out their friend. Before I went to see her, I removed a small Gideon Personal Worker's Testament from my suitcase, for the purpose of presenting it to the unbelieving flight attendant.

I found her in the first class section of the plane reading a newspaper, and I asked her if I could share some things with her. She knew that her flying partners were believers, and I told her that the members of my cockpit crew were all believers as well. I inquired about her familiarity with the term "Christian rapture." She had heard the term, but she knew very little about it. I proceeded to explain it to her. After giving her an overview of this term, I continued to tell her that if the rapture were to take place today, she would be the only crew member left on the airplane. Her curiosity was stirred as I told her how she could join us in being caught up in this wonderful event. She did not made a decision to accept Christ at that time, but she excitedly took the testament I gave her, with the promise to examine the contents. God has given me some wonderful and unusual opportunities to pass out His word.

As I ponder some of these events, I recall a vivid story that illustrates the importance of seeing Christ in all aspects of our life. Nearly everyone is familiar with one of the most famous paintings ever: *The Last Supper* by Leonardo da Vinci. This painting portrayed Christ and the twelve apostles at the table.

A great number of students of art history believe that the painting, when first produced, was somewhat different from the edition which we now see. Originally, it is believed, there was a beautiful lace border on the tablecloth. When the work was completed, da Vinci invited a group of art students to view his masterpiece, and they were enormously impressed by the delicate design of that lacework. They studied it intensely and praised it highly.

When the artist saw the reaction of these young men, he took a brush in hand and dipped it in oil. He made a few long strokes across the canvas, removing the lace by making the table cloth plain. With uncontainable emotion, he shouted, "Now, you fools, look at the face of Christ!" We are too often caught up with the fluff and fancy aspects of life, and we lose our focus. We must readjust our priorities and concentrate on the things that really matter. We must take our eyes off the beautiful lace and focus on Christ.

I was passing out New Testaments for Gideons International at the University of Santa Clara in Santa Clara, California, when one student came by and I offered him a Bible. He said that he didn't want one but he would accept something that would help him on his test the next day. I informed him that this little book was full of things that would help him on his test. He paused with a curious look on his face and asked, "Are you serious?" I told him that I was as serious as I could be. He was now ready to be shown. When he asked how this little book could help, I thought, "What have I gotten myself into, and how am I going to explain this?" Friends, God is faithful and just to provide the answers we need when we are promoting His Kingdom. Without knowing how the Holy Spirit was going to rescue me, I said to the young student, "Let's turn to parts of Scripture that provide help for college exams." He said he was all ears.

In the front of the New Testament printed for the Gideons, there is a list of many issues pertaining to the spiritual condition of the human race. As I opened the book to the first page, one thumb fell just below the word "anxious." The other thumb rested just below the word "discouraged." I did not intentionally place them there. I told the young man that I could see he was anxious about the lack of preparation for the exam, and together we looked at what God's Word says about being anxious. The very first passage we turned to was Psalm 46. The first verse of that chapter says, "*God is our refuge and strength, a very present help in time of trouble.*" We briefly looked also at Matthew 6:19-34 and Philippians 4:6.

We turned our attention to the word "discouraged." I said that it appeared to me that he was discouraged because he may not have prepared properly for the exam. I said, "Let's see what the Bible says about being discouraged." One of the verses we looked at was Psalm 55:22(KJV). "*Cast thy burden on the Lord, and He shall sustain thee: He shall never suffer the righteous to be moved.*" This young student told me that he had no idea that God's Word contained things such as this. He gladly accepted my offer to give him this Bible with his promise to examine it in greater detail.

A Jewish student came by and said that he didn't want a New Testament, but that he would accept something with the Old Testament in it. It just so happened that this little testament also contained the books of Psalms and Proverbs, which is from the Old Testament. I invited the young man to take a look at the cover of the book, and said, "How 'Old Testament' can you get?" He then said that he would gladly accept the book under those circumstances. As he left with the copy of God's Word, I reminded him of the important message contained in the rest of the book he held in his hand. I will not be surprised when I get to Heaven to see some of these people come up to me and say thanks for taking the time to give them that book. I plant the seed, but I leave it up to God to give the increase.

I have had some opportunities to visit juvenile halls and jails, and I even went to San Quentin Prison to speak with people who had made some poor choices in life. In the juvenile halls, I have spoken to kids who were locked up for crimes ranging from simple theft to murder. The same is true for the jails I visited. There were times during my visits to San Quentin Prison that I had the opportunity to talk to prisoners who had spent time on death row. Having been behind some of those walls, I thank God for allowing me to make choices that kept me outside those walls. I heard several remarkable stories from some of the inmates.

Now, let me return to my times of sharing with my crewmembers. As I gave my testimony to one of my fellow cockpit crewmembers, he asked me this question. "Why do people have to get to such a low point in order to accept the gift of salvation?" I told him that you didn't have to get to that point. You have to recognize the fact that you are a sinner and believe in Jesus Christ, and then all you have to do is repent and believe. I said that when you are down and out, it is easier to recognize your need of a Savior. When everything is going well, we seem to think we can do everything on our own. As a result, we continue running our life without the aid of Jesus Christ.

I once flew with a Jewish crew member who claimed to be an atheist. I told him that I had never met anyone of that faith who made this claim, and asked if he would be offended if I could prove to him that he was not an atheist. Since we were going to be cruising at about thirty-five thousand feet for the next few hours, he told me to go ahead and give it a try. In the next couple of hours, I not only proved to him that he was not what he claimed, but he revealed to me a number of remarkable events. He said that he did not believe the Bible as it was presented, but that it was a compilation of myths and poetry. He was fascinated by some of the stories it contained, but only if he could think of them in this context.

Our conversation progressed, and he eventually told me about having two young daughters at home. Here was a man who had claimed to be an atheist and didn't believe the Bible to be true. With this in my mind, I was now about to hear something that I could not have imagined. He said that every night when he was home, he read the Bible to his young children. I told him that his statement puzzled me because of what he had been telling me. He then made this amazing statement. "Archie, I found out a long time ago that I can be wrong. I want my girls to be exposed to the Bible, and they can make up their own mind as to its validity when they are older." He told me that he never told his children about the fact that he doubted the validity of this precious book.

I told him that the Holy Spirit was at work in his life, and that the Bible would one day become alive to him in a way he would never dream. I pray that this individual will one day have the truth revealed to him in such a way that he can't deny it. He was truly a pleasure to work with because he already possessed many wonderful characteristics of a Christian. I believe God was at work in his life.

I occasionally worked on Sundays, and as I gained more seniority in my career, I had more control over the days I worked and the trips I flew. I tried to arrange my schedule to have Sundays free in order to attend church with my wife and children. This next story is about a time when I was working on the Lord's Day.

The flight was from San Francisco to Atlanta on a Sunday with a little over four hours in the air. We completed our taxi and takeoff duties, and after climbing to our cruise altitude, I placed the airplane on automatic pilot. My copilot and I monitored our flight progress and stayed continually aware of the mechanical soundness of the plane. With many of our duties now done, we sat back and continued a discussion we began earlier about how numerous aspects of the Bible apply to every aspect of our daily lives. This crew member was a fairly new believer, so he was in agreement with most everything I had to say.

Since our discussion was centered on such an important subject, the four hour flight seemed as though it was reduced to only a few minutes.

Once in Atlanta, we departed the plane to go to the hotel. As we walked through the concourse, hundreds of people hurried to their flights. I remarked to my first officer that this was Sunday and these people should be in church. My flying partner responded by saying, "Don't look at me, I've been in church for the past four hours."

I sometimes visited San Jose Juvenile Hall on a day between trips. The ages of the boys ranged from eight to eighteen. They were detained for many different reasons, ranging from petty theft to murder. On this day, I was there with two other people on behalf of Gideons International to pass out Scripture to these young people. I had allotted about two hours for this assignment. The chaplain escorted us from cell block to cell block, and we offered a New Testament to anyone who would accept it. Many of the children were glad to receive this valuable gift while others refused.

After spending two hours at this facility, I said my goodbyes to the men who came with me. We walked down a long hallway and approached the point where we would split up. They planned to stay longer. As I started to leave the building, I felt the urge to stay. I readjusted my plans by a few minutes, and once I had decided to stay, was informed that we were going to visit a cell block that wasn't available earlier. This group had been away for their exercise break when we were there earlier. My decision to stay turned out to be a wonderful blessing for me, because it turned out to be a divine appointment.

As we entered the first cell block, a sixteen year old boy yelled, "I need to speak with someone." His cell was the first one we encountered. I went over to see what he needed, and he said he wanted to speak with someone about the answers in the little book we were passing out.

We seldom had the opportunity to go inside the cells of these kids, but the chaplain made an exception in this case and let me go inside. I sat beside this lad on the edge of his bed, and we began our little chat. He looked as if he belonged to the meanest gang you could imagine. He shared with me about how he had disappointed his mom and dad, and that he was genuinely sorry for his past actions. He had hoped he could eventually make things right with the people he had hurt. With that information, I shared the Gospel with him, and he wanted to pray the sinner's prayer with me. We kneeled by his bed, and he repeated the words of the prayer with me. When we stood to our feet, there was no question in my mind that he was sincere. I knew that I would one day see this young man again when I got to Heaven.

I didn't make my original appointment on time, but the appointment that replaced it was of so much greater importance. We have to be willing to obey that inner voice that calls, especially when we determine it is the voice of the Holy Spirit.

Sowing the Seed

I looked around and I saw there were many needs,
So today I went out to sow some seeds.
As I made preparation to start spreading them around,
I found that there was no abundance of good ground.

The ground was hard, it had rocks, it was filled with bushes;
I hadn't even started sowing, but I already had excuses.
So I asked, "Why waste my time on ground this hard?"
I'll just go back and sit in my own backyard.

Then I realized that the ground is a picture of the heart,
To receive the seed, it needs to be softened, at times broken apart.
How do I, such a lowly soul, carry out this mighty task?
The Lord said, "I'll show you how." All you have to do is ask.

So I left my backyard and went out again,
I wanted to make a difference in a world filled with sin.
Things are different now; I feel I can make a difference somehow.
God said, "Just prepare the ground with my diamond tipped plow."

A. B. Lawson, May 6, 1995

Cockpit Counseling

I was blessed in my many years of flying with the opportunity to share my faith in Jesus Christ with others. In many of those situations, I listened to the joys and heartaches of my fellow crewmembers. One day I might have shared with a mechanic, and the next day an agent, flight attendant, copilot, baggage handler, or someone who cleans and services the airplane. If we prepare our hearts to share Christ with others, God will provide the opportunities. Even though I prayed for open doors, I didn't make it a practice to push my religious beliefs on others. I tried to be aware of when that door was open. Here is just an example of looking and responding to the right time and place. One of my copilots, not knowing anything about my religious persuasion, asked me why Gideon Bibles are placed in hotel rooms. I prayed silently, "Thank you, Lord for the open door." As a member of Gideons International, I happened to know the answer. I filled him in on the reasons why the Bible is placed in hotels and other places, and I told him about God's plan of salvation. He didn't accept God's gift in the cockpit, but I thank God that the seed was planted. From that time forward, this man had no excuse for rejecting Jesus Christ.

On December 21, 2000, I flew all night from Anchorage to Salt Lake City. When the flight attendant in charge made her pre-departure announcement, she said, "Turn down the lights so it will enhance the beauty of your flight attendants." I thought about this for a moment, and I compared her statement to why the world prefers to live in darkness. It hides the ugliness of sin. My desire is to encourage people to turn up the light of God's Word to expose the elements of unrighteousness in their lives. I believe this will turn much of their sorrow into joy. This is one of many reasons why I feel it is important to share the Gospel of Jesus Christ with others.

The first opportunity to share my faith, after becoming a Christian, was in the summer of 1969. I told the captain on a flight that I had accepted Christ since I had last flown with him. His response was somewhat similar to what you might expect from someone who is outside the Kingdom of God. He said, "Archie, I have known many people over the years who have said something like what you have just told me. In my experience, I have seen many of those use their religion as a crutch." He continued, "Don't use your claim of your religious beliefs as an excuse to get you out of any trouble you might face." I told him that if I must use a crutch along with my newfound belief, I was more than happy to be a cripple, because I now possessed a peace far greater than anything I had ever experienced without the "crutch." This led to a wonderful time of sharing, and I was greatly encouraged to know that someone listened as I shared about the most important decision I have ever made. I had many opportunities to share because people saw what I was like before my conversion, and now they saw some of the changes in my life after my conversion.

I was invited to many parties on layovers before I accepted Christ. One reason for this was because I was rather good at telling "dirty jokes." Although I didn't always drink at the parties, I provided entertainment for those who did. Some called me the "life of the party." Once I gave my life to Christ, the first thing that was taken away was my ungodly language. This transformation was a problem for some because I took away some of their enjoyment. This was not a conscious decision on my part, but I simply could not be a party to this behavior anymore. The change in my behavior was a dramatic milestone in my life. It was proof that God was clearly working in my life and I was aware of what He was doing. As the unclean jokes disappeared, many would ask, "Archie, what has come over you? Did you get religion?" When you have a captive audience at thirty-five thousand feet, comments like this open doors for telling others about what God can do for them. I firmly believe God used my past to give me an occasion to bring glory to Him.

A few months later, I flew an all night flight to Hawaii. The passengers had been served and many were now asleep. One of the flight attendants came to the cockpit and shared the dark areas of her life. She was having difficulty in her marriage and her favorite brother had committed suicide within the past couple of weeks. I had just become a Christian a few months earlier, and I did not claim to be any kind of reliable authority on matters such as this. But in times like this, my faith came alive because God would usually give me comforting words to say to the ones who opened the gloomy corridors of their existence. This young lady wanted to know how life could continue when she had all of these things happening to her at once. She had seen changes in my life, and she desperately looked for some degree of peace in her own life. I shared with her that the only One who could calm the storm of her life was Jesus Christ. All she had to do was give her life to Him and turn her cares over to the only One who could restore her peace of mind. I made it perfectly clear that her problems would not immediately disappear, but I told her about a peace that the world cannot understand. She did not turn her life over to Christ in the cockpit, but she left with the assurance that there was still hope for her. I saw calmness in her as she returned to her duties. I don't know what decisions she made after our conversation, but I know I gave her advice on how she could start on that road to a new life.

Other flight attendants would often come up and tell me about their involvement with married men. Some of these women were married themselves. Again, I was no super authority on the Word of God, but He often gave me the scripture I needed to show others the error of their ways. I clarified that it was not what I thought that mattered, but it was what God said that was important. The eyes of many of these people were blinded to the negative impact they had on their own lives. Some listened and heeded what I said. Others continued in their manner of life and many homes were destroyed as a result of this behavior. Satan is powerful when it comes to convincing others about how much fun they can have, but he never tells them that it is just for a short season.

I was on a trip in late 1969 that included an overnight stay in Denver. On the layover, the captain and I, along with two flight attendants, went to dinner together. The ladies were staying at another hotel quite a distance from where we were located. After dinner, the captain had made arrangements to spend the night with one of our dinner companions without telling me. The second lady asked me if she could spend the night with me, and I told her that she could not. Since I had just recently turned my life over to Christ, I knew this was not acceptable in His sight. I was not trying to be rude, and I did not have the intention of making it look like I was better than anyone else. After I turned down the invitation, the first woman turned down the captain and they both returned to their respective hotels. The captain invited me to his room. I knew he was not very pleased with the fact that his plans were changed because of my decision. He said, "Archie, I want to know what makes you tick." He knew about my newfound faith, but he had seen others who made this claim with no dramatic change in their lifestyle. His question opened a door for me to tell him about how my dependence on Christ was altering the way I lived my life. I wish I could say that this story had a happy ending. Unfortunately, he continued in his worldly ways and his marriage ended in a tragic divorce a few months later. I don't know about his decisions later in life, but I did give him a clear presentation of the Gospel. The choice was then up to him.

In the late 1960s and early 1970s, there was a new term that received a great amount of attention. The term was a concept called "open marriage." This craze had devastating effects on marriages around the country. I flew with a fellow crewmember who was a strong proponent of this new practice. When I asked him about how his wife felt about open marriage, he said that he did his best to convince her about the benefits of this lifestyle. I told him that he was treading in dangerous territory, and that this behavior went against God's plan for a successful marriage. He replied that the Bible did not have an answer for everything. I warned him that he was only fooling himself and justifying his own behavior. I knew this individual very well and he had convinced himself that it was okay for him. However, he would be greatly disturbed if his wife ever took up this practice because he didn't believe that she would ever involve herself in "open marriage."

A few months later, I was at the airport in San Francisco with Jody when this young man approached me and said, "Archie, I wish I had listened to what you told me about my lifestyle." We had lunch together, and he went on to tell me that his wife finally had taken him up on this new concept. She had fallen in love with a community college professor and had asked for a divorce. When he learned about her indiscretions, he was so devastated that he attempted suicide. A nurse in the emergency room saw that he was a pilot from his wallet identification and called his employer. He lived through the ordeal, but he was fired and was never able to regain his job. We can make a lot of excuses about how it is acceptable to live a sinful life, but the road of life is littered with broken lives that are paying the consequences for their actions. There is a high price for enjoying the pleasure of sin for a fleeting moment.

Over the years I have worked with people who searched for the elusive dream of happiness. It seems that everyone had an answer to where it could be found. Some tried to find it in a bottle. Others searched for the dream in some sort of sexual fulfillment or other forms of "entertainment." A flight attendant came to the cockpit on one of my trips and asked if she could sit with us for awhile. I told her that she could stay as long as she wanted as long as it didn't interfere with her cabin duties. She wanted to know what the copilot and I were discussing, and I told her we were talking about certain religious subjects. She said that our topics sounded more uplifting than those at the back of the airplane. While the passengers watched the movie, the other flight attendants discussed how many sexual partners they had over the course of their careers. The numbers ranged from 5 to 32, and she said it was depressing.

I asked her what she thought of all that exposure. She felt that it was necessary to get to know the person you were going to marry. She was thirty-seven years old and had been married five times. And with five failed marriages, she still bought into the concept of experimenting sexually before marriage to "get to know" your partner. I attempted to convince her to reevaluate her viewpoint. Her theory had proven to be unsuccessful in finding true love. She admitted that her way might have been flawed, but she was not ready to change her lifestyle and move forward. I shared with her how Jesus Christ could meet all the needs of her life and give her the kind of love that would always remain true. Regardless of what she had done in the past, God was ready to accept her just as she was. All she had to do was ask God to forgive her and invite Him into her life.

Another flight attendant came to the cockpit on a different trip and told me about her wedding plans. She was radiant and excited about her upcoming marriage, and she wanted to tell me about the wedding vows she had written with her fiancé. I told her I would be delighted to listen. As she read the vows, my delight turned to sadness. In her words, she substituted the phrase, "As long as we both shall love" for "As long as we both shall live." When she came to this line, I interrupted her to ask if I had misunderstood what she said, and she said that I had heard correctly. I asked who devised that idea and she replied that her fiancé had changed the phrase. He had been married twice before, and he didn't want a traditional ceremony. I warned her to postpone the wedding until she knew more about this man, and she replied that this was not feasible because the plans were made. I believed it was better to change plans than make a mistake she would regret for the rest of her life. She thanked me for my concern, but felt the wedding plans had progressed too far to back out.

I saw this young lady again about a month later. She did not display the same countenance that she demonstrated in the cockpit earlier in the month. She looked as if the world were coming to an end. Her demeanor was similar to someone who had recently lost a dear friend in death. The first thing I thought about was her recent wedding. After only three weeks of marriage, her new husband was already cheating on her. The "as long as we both shall love" was not very long for him, and he was once again ready to conquer the heart of another lonely target.

Another young flight attendant approached the cockpit to tell me that she was getting married. I had seen her about a month before, and she was not dating anyone. I asked her about this, and she said that she had met the man she was going to marry only two weeks before. As she relayed her experience about how she met this man, my heart began to ache for her. She had picked him up as a hitchhiker in Santa Cruz, California, and he persuaded her that she was the best thing that had ever happened to him. In the process of being swept off her feet, she had neglected to learn important details about the life of this man. She thought it was unnecessary to learn more about him. She was in love! I told her about the kind of love she needed to discover, instead of the kind that caused her to be blind to things that could have a devastating effect on her life. She disregarded my counsel and said that I didn't understand the relationship she had with her soul partner. She said that I didn't fully understand "true love." I tried in vain to convince her that God had a more perfect plan for her life.

As with the previous young lady, I saw her a few weeks after her wedding. The man had moved into her apartment, and he was letting her support him. He said there was no need for him to work as long as she was making enough money to support them. It didn't take long for her to find out that he was filling his days with more than a regular day job. He would entertain other women in their apartment while she was away on trips. This was another tragic mistake that quickly ended in divorce, leaving scars that will remain for a lifetime.

> A man who wanders from the way of understanding will rest in the assembly of the dead.
>
> Proverbs 21:16

One day I shared stories with a member of the Catholic Church, and we exchanged our views on the importance of following rules and regulations. Although he had been part of this church all of his life, he possessed a limited knowledge of the Scriptures. I asked this young man what he considered to be most important in his life. He said that the biggest thing he looked forward to was going out for a drink with his priest. I expressed my feelings on this subject without making a judgment. He then asked the question, “Are you one of those “born againers?” I answered that I considered myself to be in that category. When I made this claim he, said, “Tell me what it means to be born again.” I silently prayed, “Thank you, Lord, for giving me this opportunity.” This discussion opened the door for me to give a full and complete explanation of the Gospel message. In all of his years of attendance at his church, he had never heard the good news presented in this manner. He did not make a decision in the cockpit, but he left that conversation knowing exactly what choice he must make.

I have heard people say that they were undecided about making a choice about Jesus Christ. They were often surprised to find out that they really were not undecided. They had simply decided to not consider the matter of their salvation at that time. Their decision had been made up to this point. They were passing up a chance to enjoy the peace that surpasses all understanding.

Cockpit counseling involved listening to hundreds of stories of how lives were turned upside down. People shared the hurt in their lives at any given moment. I often heard accounts such as one about a child who had been arrested for drug possession, or about a teenager who had gotten pregnant. Some shared that they just found out that their spouse, or other loved one, had been told they had cancer. The stories my coworkers shared with me over the years included just about every aspect of life, and this opened many doors for me to share about the only One who could provide true comfort. God is the ultimate source of comfort for anyone who has a burden that seems impossible to bear on their own. Rarely did I go through a day when I failed to encounter someone who needed to hear about the One who can bring healing to their troubled life.

In July of 2000, I was flying from Seattle to Dallas. One of my cabin crew members was a fifty year old lady with a suitcase full of concerns. Her husband was recovering from a debilitating stroke, and her son had committed suicide one year earlier. Her daughter had just been diagnosed with an aneurism in the brain. She told me that God had been successful in getting her attention, and He was close to doing the same with her husband. This troubled woman was a believer in Jesus Christ, but she was married to a non-Christian. While she was taking a break from her duties, I provided a sympathetic ear. I did my best, with God's help, to extend an arm of compassion to her. I gave her my phone number with the instructions to call if she ever needed to talk to someone. After we landed in Dallas, she went home to be with her family, and she thanked me for just listening and giving her the assurance that I would pray for her and her family.

In Orlando, Florida, I briefed the flight attendant in charge for our flight to Dallas. Numerous thunderstorms surrounded our route of flight. Since we expected to encounter turbulence, I wanted my crewmembers to be aware that there would be times when I would not permit them to serve. They were to take their seats if I gave that command during the flight. Nothing out of the ordinary occurred during this briefing, but one young flight attendant had the perception that the flight was going to be much rougher than normal. My briefing was intended to let the cabin crew members know that I was concerned about their safety.

Shortly after takeoff, this young lady came to the cockpit with tears in her eyes. She had become so frightened that she thought the airplane was going to crash. The air was not so terribly rough, but she had created a situation in her mind that was far worse than what we encountered. Once at our cruising altitude, I invited her to sit down to talk. She relayed to me that she had become extremely fearful of flying in recent months, and she was close to the point of quitting her job. She said that she enjoyed working with the passengers and crew, but that she was overwhelmed with the fear that had engulfed her. I had the opportunity to tell her about the One who could help her deal with her fears. I gave her a copy of God's Word with some selected verses she could read when she got to the hotel. She was calmed to the point where she returned to the cabin to resume her duties in an effective manner. When we landed in Dallas, she came to the cockpit and gave me a great big hug and said, "Thanks for being a daddy to me." The world is full of people who need someone to be their "daddy."

On one occasion, I prepared to execute an approach into Salt Lake City. I removed my approach plate from my manual and it was obvious that it was highly stained. I explained to my copilot that it contained marks from something you would never see on the chart of an obedient Mormon. He was a member of that church and the stains were from spilled coffee. This is when I was given the opportunity to explain the "parable of the stained approach plate."

The stains represent sin in our lives. In this instance, the chart was clearly readable, but approaching the limit of its usefulness. When we examine our lives and see the obvious blemish, it is time to go to the One who created us and request that the spot be removed.

> If we confess our sins, He is faithful and just to forgive us our sins and to cleanse us from all unrighteousness.
>
> 1 John 1:9

I informed my first officer that I would make the call when in Salt Lake City, and request a replacement chart. To remove the stains of sin from my life, I have to confess those imperfections to God. This is when my approach plate is wiped clean, and I am ready for my next arrival with a clean slate. God gives us the promise that all the blots of sin will be removed.

Jesus and the Chief Pilate

I admit that I didn't get the best grades in English while in college, but I did not misspell a word in the title to this chapter. As you read on, this will become apparent.

When I began a series of new trips with someone with whom I had not flown, I was required to brief my flying partner about everything I expected from him. In my opinion, this was unnecessary for pilots to do because our job description was spelled out rather well by the company and the FAA. Different people have certain ways they like to do things, so it is important that my fellow cockpit crewmember knows what he can expect of me and what I can expect from him.

Upon beginning one such series of flights, the first officer and I exchanged introductions and I gave my briefing. When I had completed the task, I asked if he had questions about anything I may have failed to cover. He said he had no questions, but he stated that he did not discuss religion in the cockpit. No one had ever said that to me at the end of a briefing. I have been told by others that religion is not something they want to discuss, and I honored their request.

About two hours into our flight, the first officer said something that had religious connotations. This prompted me to ask him why he didn't discuss religion in the cockpit. He then told me this story. He said, "I once flew with a guy whose opinions were so far removed from mine that I became very angry. I considered reporting the captain to the chief pilot for harassment." I asked him who that person was, and he said that it was me. I flew the Boeing 767 and 757 for nearly eleven years, and I rarely flew with the same pilot more than a handful of times. Years might pass between the times I flew with the same individual. I usually have a good memory for faces, but I did not remember having flown with this person.

I told him that I didn't remember our trip together, and I had no recollection of our misunderstanding. As I tried to recall our time together, I said that if anyone ever told me they had a problem with my discussing religion, I always honored that. I also made it very clear to this pilot that talking about my religious beliefs was more important to me than my job as an airline captain. I tried to take advantage of the opportunities given to me to share my beliefs with others. As I looked for those precious times, I tried to be sensitive to others. I never made a habit of ramming those beliefs down someone's throat. Something had to be said that prompted me to share. God has provided many wonderful situations for telling others about the love of Jesus Christ.

My copilot apologized for his abruptness earlier in our conversation. He understood that he must have said something during our time together many years ago that motivated me to talk about my religious beliefs. He said that we had flown together one time since, and he made it a point to stay away from any subject that hinted of religion. In that short time together, he said that I didn't bring up the topic at all. This put us back on rather sound footing, and we had some good discussions once again without any trace of bitterness.

After we had cleared the air a little, I told him that I would have felt in good company if he had reported me to the chief pilot for harassment. He said, "What do you mean by that?" I said, "Jesus Christ was once sent to the Chief Pilate, and when He was sent, He paid for my sins at Calvary." When I feel as if I'm in His presence in any situation, I praise His holy name. He laughed and had trouble containing himself when I made that statement.

This copilot had a sense of humor, which he demonstrated as he asked me this question. "Archie, did you know that God drove a Chrysler product?" I told him that I was not aware of where this was mentioned in the

Bible. He reminded me that the Book says that once "God left in a fury." I think he was stretching somewhat to come up with this interpretation, but I let him have his fun.

Now that you have heard about Jesus and the Chief Pilate, let me tell you some of the conversation we had before discussing this. One thing that I included in my briefing was that I did not permit unauthorized reading in the cockpit because it was company policy. Authorized reading was perusing material that directly pertained to the flight, such as airplane manuals, company policies and procedures, and approach plates. Unauthorized materials were newspapers, magazines, books, or other things that have no connection with the flight. It was not safer to read an aircraft manual than a newspaper, but it was company policy. Some interpreted this policy differently, so there were people who were upset that they couldn't read the *USA Today* on one of my flights. My policy was to follow the rules explicitly, unless I had a good reason to deviate for safety. That did not sit well with some copilots.

My cockpit crewmember informed me that I needed to be careful because some people talked about me behind my back. I asked what they said. He said that another captain told him that I was a big hypocrite. He said I would tell a copilot that he couldn't read his magazine or newspaper, and then I would pull out my Bible and read it. I said to him, "In your opinion of me, do you think that I would tell you not to read unauthorized material and then pull out a Bible to read?" He said he had a problem with what was being said about me because he didn't believe I would do that. I told him that I don't mind if anyone tells the truth about me, but I hate to see people hurt themselves by telling lies. I asked, "Why do you suppose someone would say those things about me?" He said that he either had me confused with someone else, or that he just didn't like me because of my restrictions. He told me about another man with whom he had flown that does read the

Bible in the cockpit. He believed that it was possible that the pilot could have thought I was reading the Bible, when in fact I was quoting from it because I knew the good Book better than most people he knew. He had paid me a wonderful compliment with that remark, and I was very humbled.

Landing for Turkeys

When a flight is completed, I don't want people to get off the airplane and say, "That was a real turkey." I never had a passenger say that about one of my landings, but I'm sure there were a few times when it could have been said. In the dictionary, one of the meanings for turkey (slang) refers to something that fails or flops. Some people refer to a bad theatrical production as a real "turkey," or it could be used for almost anything that didn't turn out quite the way it was expected. This term has been used in many ways, and I don't mean the main food attraction for Thanksgiving. Here is my story about landing for turkeys.

In the county where I was raised, there were organized turkey shoots near Thanksgiving each year. Each participant would pay a small entry fee for a chance to win a turkey. One so fortunate to hit the bull's eye with their rifle shot could cart home part of their holiday feast. This type of activity was handed down from the frontiersmen of early America. Weekend target shooting, sometimes called turkey shoots, was a popular pastime. People stood in line for the chance to take home the big bird. My story about winning turkeys has a slightly different twist from this type of shooting because we aimed for the runway.

Since my earliest days of flying, I have observed how a few pilots placed bets on landings that included variations of how one would win the bet. One way people wagered was on who could make the smoothest landing. Another was determined by placing a chalk mark on the nose wheel tire. The pilots involved in the bet would pick a point on the compass such as north, south, east, or west. When the airplane came to a stop, the winner picked the point closest to the terminal. The prize for victory ranged from the loser having to buy drinks to a small monetary reward. I never participated in these bets, though I was invited to many at different times. I'm sure that some felt I thought myself too good for their little game. This is what led me to introduce a new twist to this tradition.

After much arm twisting, I finally agreed to a wager, but the winner would not be the beneficiary. The loser had to deliver a turkey to a needy family. Here is how it went. When I flew an airplane that had three crew members in the cockpit, the flight engineer was given the job of grading the landing. I established criteria that would be considered in the judging process. The landing received the most weight, but we could be penalized for failure to keep the speed within five knots of optimum for the approach. We would also have to be on an ideal glide path from five hundred feet above the ground to touchdown. In a couple of instances we allowed a passenger to judge, and when this occurred, the smoothness of the landing was all that was considered. Even though this contest was for a good cause, no one wanted to lose. I haven't met many pilots who didn't believe they were the very best airman in the world. We are a prideful bunch.

I flew as a copilot for someone who had personal financial problems, and we decided to bet on our landings. In this particular case, I won the bet. The captain was not pleased to come out the loser, but he reluctantly fulfilled his obligation to buy the turkey for a needy family. He was so devastated by this loss that he never joined me again in this venture. Most of the people who played this game felt really good about sharing what they had with others. For some, this good feeling came after the initial shock of losing the bet. Buying the turkey was not the issue, but rather the thought that someone else had performed better than them. The person who showed no joy in having to buy the turkey declared bankruptcy a short time later. It was not because he lost the bet.

Some people have said that pilots are cheap and try to get the best price on anything they buy. A pilot who rode on my cockpit jump seat once told me this little tidbit about cheap pilots. When someone requested to ride in the cockpit, we asked a number of questions and they had to show proper identification. The majority of people who used this benefit were fellow pilots. This pilot said he had a sure way of determining whether or not the person making the request was a pilot. He said he would ask the person if they had ever bought a newspaper at the hotel on a layover. If they answered "yes," he knew they couldn't be what they claimed. He said that he never knew a pilot who had bought a paper at a hotel when he could get one free. I know he was trying to be funny, but he was also serious. There may be a spot of truth to what he said. I have read many discarded papers in hotel lobbies. Why pay for it when you can get it free?

I was the copilot for a person with whom I really enjoyed flying, and whom I still see at reunions. We made the decision to make a bet on our landings. After we had both made our landings, the flight engineer declared a tie. This meant that the loser on the next bet would have to buy two turkeys. We each made our takeoffs and landings, and again came out even. The stakes grew higher because the loser would now have to deliver three birds to three different families. The next takeoff and landing belonged to the captain. His approach and landing were executed almost perfectly, and I did not feel I had a chance in the world of topping him. We were flying into Reno, Nevada, and I was gladly prepared to go buy the food.

The next departure took us from Reno to Salt Lake City where I flew the plane. I knew what I would do on my days off, and I felt that this was perfectly okay because some needy families would eat well for a couple of days. Before I could make my landing in Salt Lake, the captain said, "Archie, the contest is over. I have thought about what you are doing, and have to say that I like it. I want to buy those turkeys myself." I had never seen anyone do this and it was a very unselfish act on his part. Once I landed in Salt Lake City, I became ill and had to fly back home to California. The captain and flight engineer continued on the trip with another first officer. As I flew back home that afternoon, I left with a warm feeling of gratitude for what this pilot had done.

When I arrived home that evening, I received another surprise. The captain called me from his hotel in San Diego and said, "Archie, those people need something to go with those turkeys. I'm leaving a one hundred dollar bill in your mailbox at work for that purpose." When I felt better, Jody, David, Peter, and I went to the market and purchased the three turkeys with all the trimmings. The families who received the food were very grateful, and it was all because of one man's generosity who didn't mind losing for a good cause. In losing, he was actually the victor.

I hadn't seen this friend for several months when Jody and I saw him at our company Christmas party at our chief pilot's house. I thanked him once again for his gesture. As we talked, he reached into his wallet and handed me another one hundred dollar bill. He asked if I would take it and buy more food for other needy families. His act of kindness was a marvelous blessing for me, and I know from his expression that he received a blessing from doing something worthwhile for those less fortunate. I am not a person who gambles, but I had no problem with landing for turkeys. In most cases, even the loser came out feeling good for his chance to help others.

Low Visibility Approach

A low visibility approach in an airplane parallels some of the spiritual choices that we make. In life there are times when we aren't completely clear about what will happen, and we live with too many "what ifs." "What will happen if I do this or that?" or "I avoid doing certain things because I am afraid of the outcome." We must listen to warnings about doing certain things, but at times we respond to the wrong message rather than using reason to make choices. In executing a low visibility approach, I had to consider a lot of "what ifs," but I would not permit the questions to keep me from doing it.

Under certain meteorological conditions we were required to execute low visibility approaches. The approach category that permitted us to land at an airport with the lowest visibility was a category III approach. This allowed us to land with visibility as low as three hundred feet. That seems like a long way, but when traveling at speeds in the neighborhood of one hundred and fifty miles per hour, it goes by rather quickly. We had to adhere to very strict guidelines when we performed an approach under these conditions. Not only did the equipment on the airplane have to be working properly, but certain airport criteria also had to be met. If we performed a category III approach with three hundred feet of visibility, we would sometimes land without ever seeing the runway. The approach was fully completed by the use of the autopilot system. On the Boeing 767 and Boeing 757 aircraft, we had three autopilots.

The job of the pilot was to program the computer and closely observe all aspects of the approach, making sure that we stayed within very restricted boundaries. This method of flying was very safe if we stayed within those guidelines. This approach was not an everyday happening. In eleven years of flying these airplanes with this capability, I saw only four times when it was necessary to fly approaches this restricted. We received thorough training each year, and it was great to know it was available if needed.

On one such trip, my copilot and I departed San Diego around midnight, and we were scheduled to arrive in Cincinnati early the next morning. About an hour from landing, it became apparent that the weather was rapidly deteriorating as far as the visibility was concerned. There was no need for worry because we had the ability to land under these conditions. My copilot had never seen an actual category III approach, although he had practiced many under simulated conditions. He was somewhat apprehensive regarding the outcome. I had about an hour to assure him that everything would be okay.

Why would he be a little nervous about what lies ahead? When we prepared for this type of approach, numerous options had to be considered.

- Was the equipment on the airplane working correctly?
- Were the conditions at the airport as they should be?
- Did we have the proper approach plates with which to program the computer?

All of the approaches were in the computer and we had to check that the information was correct. In addition to preparing for the approach by setting up and checking the equipment, I had to consider what would happen after I landed the airplane. Once I turned off the runway, how was I to make it to the gate if I couldn't see what was in front of me? When we flew the approach, we had to be continually cognizant of all aspects of the approach and landing.

- Was the autopilot keeping us on the centerline of the runway?

- Was all of the necessary equipment continuing to function as it should?
- Did the wind stay within crosswind limits for this approach?
- Was I prepared to take over and execute a missed approach if anything out of the ordinary happened?

These questions could cause a person to become somewhat anxious if there were uncertainties. I always had faith that the airplane would do its job, but I never had blind faith. I was prepared to terminate the approach at any time if the conditions were not right. This was a time when we could relax a little while sitting through an intense situation. However, we could never permit ourselves to become complacent.

Once we had made the necessary preparations for the approach and everything checked out fine, we proceeded to execute the maneuver. We determined that the lighting system on the runway was operational. These high intensity runway lights line the center of the runway. Even though we weren't able to see much before landing, they permitted us to keep the airplane centered on the runway after touching down. The autopilots would do the work, but we were prepared to help if needed. We were also certain that the high intensity runway turnoff lights would help us know where to exit the runway. They would direct us to a safe distance from the runway, permitting planes behind us to continue their approach and landings safely.

We slowed the jet to a safe speed, and exited the runway with the aid of these lights. Most of the time the visibility was good enough to see taxiway markings, but at times when we couldn't see what was in front of us, we could run into uncertainty. In this case, I always considered what I would do next. I made a request on the radio to my operations for a "follow-me vehicle" and someone was dispatched to lead me to my gate.

After we have completed our final approach, what will be waiting for us? Will we end up in a big black hole of utter darkness, or will our Savior wait to take us to the gate and welcome us home? If we know Jesus Christ as our personal Lord and Savior, we never have to wonder what is going to happen after we make our "final approach" in life.

A counselor at a church camp spoke of his experience with a nine year old boy who cried when the cabin lights were turned out on the first night. The little boy was asked if he was fearful of the dark. He said he was not afraid of the dark, but exclaimed, "I just don't want to be attacked by the 'killer rabbits'." Some older kids at home had told him that there were "killer rabbits" that would come out at night and attack the campers. He was concerned about the "dark hole." Jesus constantly reassured the disciples with the words, "fear not." Over and over again, we see that their fears betrayed their lack of faith. These words illustrate man's constant need to be reassured of God's presence and His comfort. A child of God can draw on this presence to find comfort and wipe away fears. Watch out for "killer rabbits!" They can destroy peace of mind at camp and throughout life.

Low visibility approaches sometimes remind me of people going through life worried about what is coming next. They seem to have no assurance that today will be any better than the day before, and they approach life with little hope of improvement in their circumstances. The reasons for this may be financial, physical, marital, or otherwise. Much of the time, the cause is due to choices they have made. I have been involved in voluntary financial counseling for a number of years, and I saw that people usually had money problems because of poor choices. Many physical problems that people have can be attributed to the choices made in their dietary habits. If a couple has marital problems, it is normally the result of choices made in regards to their mate. If I am going to be successful in the execution of a low visibility approach, I am required to make the right decisions and choices. My failure to do so will place the lives of many people in peril. When people fail to make right decisions and choices in daily life, a lot of lives may not be in jeopardy, but peace, joy, and happiness will be endangered.

On each of my flights, I carried manuals that covered almost every aspect of my job. My responsibility was to know what the manuals contained. I had to refer to some manuals continually for material such as approach plates, but much of my job was done without having to refer to a book. This is because my job demanded a thorough knowledge of what was expected of me. You might say that I needed a "head and heart" knowledge of my mission. I have found that if a person has a "head and heart" knowledge of the Bible, he is much more inclined to make the right choices. When faced with having to execute a low visibility approach in life, that individual will perform the approach and landing safely. He will not be left in a black hole of doubt upon completion of the ordeal. There may be times when it seems like the approach goes on forever, but it will always come to a satisfactory conclusion.

One such example of a low visibility approach was of a woman in the middle of a situation that she thought would never end. Her problems were mainly a result of poor financial choices, and these choices had brought her to the point of considering bankruptcy.

This lady had attended one of the financial seminars that I conducted called "God's Plan for Financial Contentment." In this seminar I covered most aspects of financial decisions. My goal was to teach people how to achieve piece of mind in financial affairs based on making godly choices. She called me one day and said she remembered something I had said in regard to declaring bankruptcy. She had just left the office of an attorney who was a specialist in this type of action, and she wanted to hear again what my thoughts were. I had stated that, with few exceptions, I was against seeking this form of protection. I invited her to my home so that my wife and I could talk with her.

She came to see us, and explained why she sought this method of protection. From a worldly point of view, it was difficult to disagree with what others were telling her. With the problems she had, there didn't seem to be any other way out. Her job paid her about seven dollars and fifty cents per hour, and this wage was not going to permit her to eliminate over twenty seven thousand dollars of debt. Where do we go for help to solve a problem such as this? First, one must be confident that nothing is impossible with God. One must shy away from saying, "That may be true for others but my problems are just too great." This individual was seeking help because she viewed her situation as hopeless. After listening to the information that she gave, here is what I proposed.

This lady held two masters degrees in various fields. Why was she trapped in a job that was paying her such a meager amount? The reason was that her self-esteem had deteriorated to the point where she could not sell herself as being of value to an employer. I outlined a plan where she could pay off her debts in about two or three years if she would agree to some stringent measures. First, I advised her to call the attorney and tell him that she was not going to seek protection under the bankruptcy law. I also told her what I thought would happen if she made this choice with the help of God. If she chose my alternative to meet the obligations for which she was responsible, it was my belief that she would begin to regain her self-esteem. She would then be able to secure a better paying job, thus enabling her to meet her financial responsibilities.

After completing our two hour discussion, Jody and I prayed with her for God's wisdom. She left with a degree of uncertainty about what she was going to do, and said she would pray about the advice I gave her. About two hours after leaving our home, she called with the results of her decision. She said that she had called the attorney and told him that she was going to make an attempt to pay off her debts.

In about two short weeks, this lady had a job that was paying her more than double her previous salary. In just over a month she was earning over three times the amount. After that contact, Jody and I were not aware of what was going on with her for over a year. We attended a meeting with FCAP (The Fellowship of Christian Airline Personnel), and saw her there. A friend had invited her to this function. We renewed our acquaintance that evening, and she proceeded to tell Jody and I the results of our recommendations. In just over a year she had fully paid off her debts. She now faced the future on sound financial footing. This could not have been done without the help of God and her willingness to alter some of the ways she made choices. God allowed her to conduct a low visibility approach, with many unknowns, and brought her flight to a successful conclusion without being left in a world of seemingly utter darkness. He can do the same for us.

There are times in our lives when it seems as though we have a pretty good approach in progress. Things may be moving along at a fairly normal pace with no major problems with which to deal. When this has been a pattern for some time, we might say, "Life is great." Suddenly, the bottom falls out and we may think, "Life is not so great." How can we deal with these ups and downs without tearing out our hair? If we set our lives on a course to please God with the way we live, it will be somewhat like the preparations I made for a low visibility approach. When everything is working right, all is wonderful. If something goes wrong, you can handle it without devastating consequences.

I made an approach to Calgary, Canada, many years ago that is somewhat similar to this. We had descended to an altitude of about ten thousand feet and the weather consisted of uneven broken clouds. In some areas the visibility was great. We saw the airport about ten to fifteen miles out, and were cleared for a visual approach. Whenever we were cleared for a visual approach, I had my instruments and radios set up so that I could transition to a full instrument approach if the weather changed. With a visual approach, the airport as well as the airplane ahead, must be kept in sight at all times. The nice thing about a visual approach is that I was able to enjoy the scenery. Combine this with the enjoyment of flying and I had an awesome mix.

As we neared the airport, there were times when we lost visual contact with the runway, if only for a second or two. It was of no concern, however, because I could change from flying visually to flying on instruments. This made flying the approach very comfortable and enjoyable. This intermittent loss of contact lasted for almost all of the approach, but we landed safely and smoothly. When we are properly prepared to face life by placing our trust in Jesus Christ, we can execute our approach to life successfully. There may be times when things look a little dim, but the brightness will return if we persevere. As you come back in the clear, you will be able to complete your low visibility approach safely.

Flying by Faith

> Now faith is the substance of things hoped for, the evidence of things not seen.
>
> Hebrews 11:1

Flying airplanes is an occupation that requires frequent exercises in faith, and I am not referring to blind faith. Blind faith dictates that I may have to act contrary to everything I know and trust. Many people say, "If I just have faith," things will work out fine. I have to have faith that the aircraft is mechanically sound, but there are procedures I must follow to assure all systems are airworthy. There is much more involved than just assuming all is well. I have to place my faith and trust in the meteorologist, the air traffic controller, the mechanic who works on the airplane, the control tower personnel, and each and every member of my crew. If I am going to offer my passengers a safe and uneventful trip to their destination, I have to work in concert with all of these different people. I placed my trust in these people, but I did not do it blindly. I also place a lot of faith and trust in the pastor of my church, but there are times when I go to the Word of God to see if he communicated the message in the right manner. Paul wrote in 2 Timothy 2:15 (KJV), "Study to shew thyself approved unto God, a workman that needeth not to be ashamed, rightly dividing the word of truth." It is important to me to know that my pastor is following the instruction that Paul gave to Timothy regarding this matter. This does not mean that I am constantly on the lookout for mistakes from my minister. I discovered many years ago that people are not perfect, and any of us can make a mistake. As an airline pilot, I did everything in my power to avoid mistakes that could cause irreparable harm to those I was privileged to serve. In the seventeenth chapter of Acts, Paul speaks about how the people from Berea would go home in the evening and examine the scriptures to see if he was telling them truth. They had faith in Paul, but not blind faith.

In the beginning phases of flight training, pilots are taught to have faith in their instruments. When a pilot is flying in poor weather conditions, he has to rely on his instruments to tell him the attitude, altitude, airspeed, and direction of the airplane. In clear weather he can fly with reference to the horizon, but he loses this reference when he is in the middle of the clouds. The primary flight instruments are the attitude indicator, airspeed indicator, compass, and altimeter. All of these are extremely important, but the attitude indicator is the aviator's single most important reference. The word "faith" can have several different meanings, but I want to now consider the Biblical definition. From Ephesians 2:8-9, we learn that faith is a condition of salvation. If we place our faith and trust in the Lord Jesus Christ, we have the assurance that salvation is complete. We can have hope that something will happen, but we still have the anxiety associated with waiting for the results. If we have faith, we can assume that we already have what is promised. Faith in the Ephesians passage requires that we rely on what Christ has promised in His Word regarding salvation. Faith in my instruments requires that I rely on the promise of the manufacturer that the instrument will work as described. If I am going to fly my aircraft without fear, I must have the belief and assurance that the instruments will function as expected. Here is a story about a highly unusual and unlikely situation regarding this faith.

In the late 1970s, I was flying from San Diego to Las Vegas. On the departure from San Diego, the weather was very cloudy. Shortly after takeoff, I was climbing to my cruise altitude, and I was in the process of making a large sweeping left hand turn over the Pacific Ocean. While in the middle of my turn, I looked over and saw that my copilot's attitude indicator was spinning. This is not a good thing, but we had three separate attitude indicators to use in case of malfunction. There is one for the captain, one for the copilot, and a standby indicator to be used in the event of a power failure. When an indicator malfunctions, a warning flag and caution light should show what is wrong so that the problem can be corrected. In this instance, we had no warning of any kind. As the attitude display continued to tumble, we were still climbing in the clouds while completing our turn. I still had my attitude reference, and the standby unit was still working. However, a few seconds later, my attitude indicator started to tumble without any warning of an impending problem. I quickly analyzed the problem and determined that the standby gyro was the one on which I must rely. As soon as I began to depend on this backup instrument, we climbed up out of the clouds. We were now in the clear, and I could keep the wings level by referencing the horizon of the earth. I could not return to San Diego because of poor weather conditions, so I made the decision to continue to Las Vegas with clear weather all the way. To my knowledge, I had never heard of a situation where both the captain's and copilot's attitude indicators failed at the same time without any warning signals. There have been situations where an aircraft lost all of the primary electrical systems, but there are various warnings in these situations. Even if all of the primary electrical systems fail, the aircraft can be safely flown with reference to the standby gyro and other necessary instruments. This is not an ideal situation, but the plane can be landed safely in these conditions. We had to try to find out what caused our problem, which was considered more serious than a complete electrical failure. We were trained to have complete faith in our instruments, as long as we don't have the warning signs to show us differently. If we had remained in the clouds, we would have been faced with an extreme dilemma because we did not know which instrument to trust.

Today, we have many choices about where we place our trust. We go to a certain church, and we are told to trust in the gods they worship. A serious illness necessitates a trip to the physician, and we are instructed to place our faith and trust in some "wonder drug." While most of the people to whom we listen give their best recommendation, it is advisable to consider their suggestions in a critical manner. By this I mean to examine choices carefully. When it comes to serving God, we must make sure we are placing our trust in the One who is trustworthy. *Jesus said to him, "I am the way, the truth, and the life. No one comes to the Father except through Me.* John 14:6 The course of treatment that the doctor prescribes may seem as if it will solve a physical problem, but with further research we discover we have a medical condition that is not compatible with a particular medication. When I utilize the Pray, Think, and Act principle, I am much more likely to make the right choice. I formulated my decision regarding my instruments based on sound knowledge from extensive training and study, and I made the right choice.

After landing in Las Vegas, I called for a mechanic to come to the airplane. When he saw that I had written up both the captain and the copilot's attitude indicator power sources in the maintenance log, he told me I was crazy because this is something that could not happen. These instruments are powered by two different sources, and it is highly improbable that there could have been a problem with both units at the same time. After I calmed the mechanic down, I told him it was his problem now, and that he must take a closer look at my write up in the log book. The airplane could not leave Las Vegas until this problem was resolved. He then went to the instrument compartment of the airplane, removed both vertical gyros, and took them to the shop where he could analyze each of them. About fifteen minutes later, he came back to me and apologized for his earlier statement. He informed me that indeed both of the gyros were defective. He commented that he didn't understand how this could happen, but admitted that the problem did exist. Both of the vertical gyros were replaced, and we continued on our flight to San Francisco.

The maintenance department continued the investigation into this incident until they discovered how both of these instruments had failed without providing any warning. In San Diego, the airplane was connected to external power at the gate. This allowed the airplane to use all of the necessary electrical systems to run air conditioning and permit people to work on the plane without the engines running. While connected to the external power, there was a power surge that welded some electrical contacts in the vertical gyros. This power surge would be classified as "bad power" and it rendered the warning systems inoperative. There was nothing to indicate any malfunction of these systems. As far as we knew, we had a perfectly safe airplane to fly to Las Vegas.

Why would I tell you this story about instruments that malfunction? Some people believe that faith is simply devotion to whatever god they happen to follow. The person might say, "I am a person of deep faith because I am a true follower of Mohammed," or Ayatollah, or some other god they happen to choose. They believe that what matters is that they are sincere. This is a huge misconception. If someone chooses to plug into the wrong power source, there will be disastrous eternal consequences.

Vertigo is a condition, although rare, that a pilot can experience. Vertigo is having a sensation of spinning around or of seeing objects nearby revolve around you. This condition can be caused by certain complex physical conditions that a person might have, or caused by an infection or inflammation of the inner ear. When a pilot experiences vertigo, he has the sensation that the airplane is doing something other than flying straight and level, and the tendency is to try and correct the situation. An experienced pilot will immediately refer to his flight instruments, and he will place his faith and trust in the indications he receives from them. When flying with a two man crew, the first thing he would do is to tell the other pilot that he is experiencing these feelings, and ask him to take control of the aircraft. There are times in everyday life when we experience certain feelings that cause us to react in a particular way, and there are times when these reactions cause us great degrees of difficulty. When those feelings arise, we should immediately refer to the instrument which is the Word of God. If we observe this procedure, we can often prevent our lives from spinning out of control. A pilot is in danger of spinning out of control with vertigo if he doesn't refer to, and follow, what his instruments are telling him.

You may have heard the term, "Flying by the seat of my pants." This was a common expression in the early days of aviation, and it is still used today by people who fly aircraft without the luxury of the latest avionics. There are many antique airplanes flying today. I had an incident in 1988 where I fully experienced this concept. I was away from flying for almost 6 years for health reasons. (You can read the full story in the chapter titled, "I Believe in Miracles.") On my first visit to a flight simulator in many years, the instructor introduced me to a scenario that I would not soon forget. I had just taken off on my first simulated flight, and immediately after leaving the runway, I had a complete electrical failure. This was a night flight and everything in the cockpit was completely black. I could not see any of my instruments. I immediately asked my copilot to get out his flashlight and shine it on my instruments. The instructor said that the flashlights did not work. He said, "You have to fly the airplane by the seat of your pants." Outside the cockpit, the simulated weather was clear and I did have the outside horizon as a reference. With this view, I could keep the wings level, but I could not see my airspeed indicator or my altimeter. I was flying under VFR (visual flight rules) conditions, and I could keep my departure airport in sight. There was no problem making turns back to the airport because I had the horizon in view, but there was a problem with maintaining the proper speed. I could do without the altimeter because I could see when I was approaching the ground. I guess you could say that I controlled my airspeed by the seat of my pants. I could detect when I was getting too slow because of the stall warnings. If the speed was too great, I could feel a buffet (vibration) in the controls. A stall warning is a device that warns the pilot before he reaches a dangerously slow airspeed. When the pilot allows the airplane to slow to a critical level, this device sounds a loud warning and the control column (yoke) vibrates. This warning device is a critical part of any aircraft. My job was to fly the airplane somewhere between buffets and stall speed.

This piece of flying was not a masterpiece, but I safely flew the plane back onto the runway without any damage. The reason that I was able to do this was because I knew what to look for in each of these situations. Because I knew what to expect, I had complete faith in my safe return to the airport. I have never spoken with another pilot who has ever been faced with a training situation such as this, but I thank God for my experience in this unusual state of affairs. It illustrates that the more of God's Word we know and understand and incorporate into our daily lives, the more likely we will be able to successfully face the difficult things that challenge our faith.

In another situation, I was required to place my faith and trust in my flying partners. In flight training at Whiting

Field outside of Pensacola, Florida, I was learning to fly in formation with other pilots. I was flying with three other students and a flight instructor who piloted the lead airplane, which totaled five aircraft. We had just completed our required training on this particular day, and it was time to head back to our base. As we were returning to the airport, the weather began to deteriorate. Since this was early in our naval flight training, we had not yet been trained to fly by instruments. We were only authorized to fly under visual flight rules. It soon became apparent that we could not fly back to the intended point of landing without flying through clouds. The instructor was a highly-trained aviator and qualified to fly in almost any kind of weather. We were warned by the instructor that we would be flying through some inclement weather. Each of us knew that we were not qualified to fly under these conditions, but he assured us that if we would remain in a tight formation and focus our attention on him, we would safely fly through the clouds. This is what we did and everyone arrived at the base without incident. If any of us had chosen to break off from our leader while in the clouds, the results would have been catastrophic.

There are times when we are faced with monumental problems, but we discover that if we place our faith and trust in God during those times of difficulty, He will bring us safely through the clouds. Peter found that he could walk on water as long as he kept his eyes on Jesus. He only started to sink when he took his focus away from Christ. If we stay close to Him and do not lose our focal point, we will gain victory. In this case, the instructor gave us the warning that made the difference, but we had to heed that warning. Throughout the Bible, God gives warnings that result in either blessings or curses. I like the expression my wife often uses, "I don't believe in luck; it is either God's blessings or challenges He allows to enhance our faith."

The United States Navy's Blue Angel flight demonstration team and the Air Force's Thunderbirds are a witness of incredible faith and trust in fellow pilots. There are some maneuvers that require pilots to place their lives in each other's hands. These pilots must perform flawlessly. Failure to do so will likely result in someone being seriously injured or even killed. These highly skilled pilots perform their beautiful shows with incredible precision, and they do it with the complete faith and trust in their fellow pilots.

A study of the eleventh chapter of Hebrews is an examination of the lives of the people mentioned in the "hall of faith." It illustrates how faith caused Abel to worship God, Enoch to walk with God, and Noah to work for God. In this chapter, God uses ordinary people to carry out His work, the only requirement being faith and trust that God will do what He promises. Faith is a condition of salvation, but is also a result of salvation. As a Christian matures, his faith will increase as he relies more and more on the One who provided salvation for him.

There are many ways to increase faith. Take a moment and consider these points.

- Listen to the Word of God and put it into practice as much as you possibly can. *"So then faith cometh by hearing, and hearing by the Word of God."* Romans 10:17 (KJV)
- Know and understand that each believer has been blessed with a measure of faith. *"For I say, through the grace given unto me, to every man that is among you, not to think of himself more highly than he ought to think; but to think soberly, according as God hath dealt to every man the measure of faith."* Romans 12:3 (KJV)
- Be obedient to God and be open to the conviction of the Holy Spirit.

- Thank God for results before you see them. Avoid complaining. *"In every thing give thanks: for this is the will of God in Christ Jesus concerning you."* 1 Thessalonians 5:18 (KJV)
- Make an attempt to spend time with people of true faith. *"He that walketh with wise men shall be wise: but a companion of fools shall be destroyed."* Proverbs 13:20 (KJV)

Final Flight

I flew my last flight for Delta Air Lines on July 19, 2001. This was two days before my 60th birthday, the mandatory retirement age required by the FAA. This age regulation applies to pilots of all major United States commercial airlines carrying passengers. I didn't have to give up flying completely, but I could not continue to be employed in my former capacity. I personally believed that this was a good rule, because it provided an opportunity for a new generation of pilots to be employed. However, if the rule did not exist, I would probably still be flying because I truly loved my job.

Unknown to me, Jody and our son David began to make plans for this flight almost a year before it took place. With my seniority, I had a choice of the days I wanted to fly as well as the destinations. For my last day of flying, I chose flights that were convenient for my family to accompany me. Jody, our sons, David and Peter, and David's wife, Rebecca, flew on my last two flights. We traveled from Salt Lake City to Sacramento and back.

My boss, Chief Pilot, Captain Joe Moran, requested to fly with me on my final mission as my copilot, and it was an honor to have him aboard. Unbeknownst to me, he was involved with some of the planning that went into these flights. After the passengers boarded the airplane, I gave my routine "Welcome Aboard" announcement and thanked them for the opportunity to serve them on my retirement flight. We took off and headed toward Sacramento, and once we leveled off at our cruising altitude, I intended to speak to the people on board once more. I switched the radio button on my audio selector panel to the public address position, which let me determine if the flight attendants were talking, or if a movie was playing. I couldn't believe what I heard! I heard myself speaking at the back of the plane. I told my copilot, and he pretended to know nothing of what was taking place. He was having a good time watching me try to figure this out.

After my voice went silent in the cabin, I found out what had been going on. Almost a year before this flight, our son David and Rebecca came home for a visit. David wanted me to sit down and tell a few of my stories while he recorded them on his video camera. I asked him why he wanted to do this, and he said that he wanted to have it for his own use. The video that David shot was the centerpiece of a seventeen minute video that David, with Jody's help, put together about my life. They played this video to the passengers on the flight to and from Sacramento. I can tell you that this was a very special moment for me. In addition to the video, Jody, with the help of the flight attendants, did a number of other things as well. The passengers seemed to enjoy the opportunity to participate.

Another surprise made my mouth drop open when I saw what transpired. Jody had taken one of the pictures of me in my uniform to a graphics shop and had them duplicate a life-size face mask. She had over one hundred eighty of these made, and gave one to each one of the passengers. On my last two flights, I was flying the Boeing 757 that seated one hundred eighty two people. Our load that day was just shy of the maximum. The masks were only passed out on the final flight segment back to Salt Lake.

On the flight from Sacramento to Salt Lake, my final flight, the video was once again shown. After the video finished playing, the flight attendants said they wanted me to come to the cabin. I left my copilot in charge, and when I stepped from the cockpit into the cabin, everyone yelled thunderous congratulations. I was invited to walk down the aisle with Jody and thank the passengers for flying with me. Before I returned to the cockpit, I was told to look back once more. When I did this, the flight attendant gave the signal, "Now." In unison, nearly one hundred eighty people raised these masks to cover their faces. Some people think that one Archie is too much. Can you imagine what it is like to look at one hundred eighty at one time? I must tell you that I was both pleased and in awe at the same time. I was beginning to wonder what was going to happen next. I only knew about a fraction of it.

When I returned to the cockpit, Joe had already begun our descent into Salt Lake. I once again took over to make my final landing. The air traffic controllers had been alerted to the fact that this was my last hoorah, and many of them came on the radio with their friendly congratulations. One of the controllers was my friend from our church, Tom Meyers. This was another very special moment.

I stopped the airplane after departing the active runway, and announced to the passengers that I was going to bring my family up to the cockpit for my last taxi to the gate. Jody, David, Rebecca, and Peter came up and we continued toward the gate. I knew that the Airport Fire Department would present me with the "Water Canon Salute," because it was done for most pilots on their retirement flights. Only poor weather conditions and an after dark arrival would prevent this little element from happening. I was met at the runway and escorted to the gate for the last time by three official white vans. The escort vehicles led me on a path that took me between two airport fire trucks that saluted me with streams of water shooting over the plane as I slowly went by. It was a dramatic scene.

An airport official presented me with a video of the "Water Canon Salute" when I arrived at the gate. This was both a sad and happy moment. As I said goodbye for the last time to the passengers as they departed, many of them congratulated me once more. Some of them held the "Archie" mask to their face as they left. I will have to say there were a couple of times when my eyes welled up with tears. When the goodbyes were finished, I gathered my suitcase and flight bag and walked off the airplane for the last time.

I exited the jetway into the gate area where about twenty-five friends yelled, "Congratulations and welcome home, Archie." Jody had arranged for them to greet me, and I gave each one a hug and thanked them for coming to see my last arrival. In the pilot's lounge, Jody had arranged for a small celebration complete with cake. Captain Joe Moran presented me with a plaque denoting my years of service and made a short speech. This is small sampling of what happened on my last day of flying.

As you ponder the story I have just told you about my last flight, I want you to think in terms of the very last flight you will take sometime in the future. There are some similarities in my final flight to the one that is awaiting you. First, we know that our final trip will one day take place. Second, there are some preparations that must be made for the event. The one thing that determines your peace and happiness in between these events is how you live your life. After you have given your life to Jesus Christ, you are now ready for your final flight. Once Jody began her preparations for my final flight, her activities reflected how much she wanted this to be a glorious event. Are we living our lives in such a way that shows that we will one day be facing the most wonderful occasion we could ever imagine? This will be our entrance into Heaven. Are your daily activities a picture of what is to come? As Christians, our lives should be patterned after Jesus. If we do this we will have a most marvelous landing on our very last journey. May God bless you in a powerful way! My party was far greater that anything I ever imagined, but it pales in comparison to what is waiting for me and you when we go home to meet our Lord and Savior Jesus Christ. I will have an incredible "View from the Cockpit."

So I said, "Oh, that I had wings like a dove!
I would fly away and be at rest."

Psalm 55:6